Tweeting the Universe

Tiny explanations of very big ideas

Marcus Chown & Govert Schilling

@MarcusChown @GovertTweets

First published in 2011
by Faber and Faber Limited
Bloomsbury House, 74–77 Great Russell Street,
London, WC1B 3DA
This paperback edition first published in 2013

Typeset by Faber and Faber Limited
Printed and bound by CPI Group (UK) Ltd, Croydon CR0 4YY

A CIP record for this book
is available from the British Library

ISBN 978–0–571–29570–8

2 4 6 8 10 9 7 5 3 1

To John Grindrod – the funniest Tweeter – for getting Marcus started on Twitter

Contents

Foreword

It all started on a small island in the Caribbean. Well, it makes a better story, doesn't it? And, well, actually, it did – sort of.

Aruba is the driest island in the Caribbean – famous for its casinos, wind-bent 'divi-divi' trees, and not much else. On 26 February 1998, an alignment of the Sun, Earth and Moon meant that, for 3 minutes and 32 seconds, the island was treated to one of nature's great spectacles when the Moon blots out the Sun in broad daylight. Marcus was in Aruba to report on the 'total eclipse' for the English magazine, *New Scientist*. And Govert was there doing the same for a Dutch weekly, *Intermediair*. Aruba is part of the former Dutch Antilles.

So, to cut a long story short, Marcus and Govert met for the first time. In fact, Govert very kindly gave Marcus a lift to the airport to get his 2 a.m. flight back to the UK.

Fast-forward to 2009 when a social networking site called Twitter, which nobody could have envisaged in 1998, was rapidly gaining popularity. Govert embraced it. And so did Marcus. Actually, that's stretching the truth. Like most people, they were deeply sceptical of Twitter. Marcus had to be persuaded to give it a go by John Grindrod, the marketing manager at his publisher, Faber, who told him it was a way to talk directly to readers.

Through Twitter, Govert and Marcus became aware of each other again. They were fellow tweeters. And, in late 2010, Govert sent Marcus an email with an interesting proposal.

Because Govert's followers had been asking him lots of questions, he'd hit on the idea of doing a weekly Twitter course on an astronomical topic every Friday evening. Govert's editor at *De Volkskrant,* a Dutch national daily newspaper, had noticed and said: Why don't you do the same for us in a weekly column? So Govert did, publishing all 15 tweets the day after they appeared online. The response from readers of the newspaper was enthusiastic, so Govert's thoughts turned to doing a book for a wider audience – in English. Which is when he thought of Marcus. Would he be interested in doing a book of tweets with him?

Marcus's immediate thought was: what a rubbish idea. No, seriously, he thought it was a very good idea. So he contacted Neil Belton, his editor at Faber, who would probably be the first to admit he is not in the least bit techy. To Marcus's surprise, Neil was very enthusiastic about the idea. Pretty soon, a contract was drawn up, and Govert and Marcus were collaborating.

Distilling entire subjects like the Big Bang theory down to a series of tweets proved a challenge, to say the least. Govert had some experience already from writing his weekly tweet column for *De Volkskrant*. The only practice Marcus had was *Solar System for iPad*, where no story he wrote – about a planet or moon or asteroid – could be longer than 275 words in order to fit on the iPad screen without the need for scrolling to another page. Two hundred and seventy five words is short but it is novel-length compared with a tweet, which has a maximum length of 140 characters.

Both Marcus and Govert soon discovered that what they had considered at the outset would be a relatively quick project was far more time-consuming. Since over-compression is synonymous with incomprehensibility, striking a balance between distilling a topic down to its essence and conveying real understanding to the reader was hard. Added to that was the difficulty of continually coming in under 140 characters. Often, it took more time to shave off a few surplus characters than it did to formulate a tweet in the first place. Marcus found himself scribbling in a notebook while walking in the park, waiting in supermarket queues and riding on the top deck of London buses. For Govert, who spent long hours at his desk, a walk in the park was but a distant dream!

Agreeing that the obvious number of topics to cover was 140, Govert and Marcus set about covering 70 each. When they finished, they swapped their material and edited each other's words. That was another time-consuming process they had failed to anticipate. But, at last, it was done. In the space of a year Marcus had gone from writing at book-length to 275 words to a mere 140 characters. Govert no longer utters a

The Sky

1. What causes a rainbow?

1665. Plague in London. To northeast, Cambridge closes. Newton, 22, unknown, treks home. Secluded for 18 months, he changes face of science.

During his 'miraculous year', when he explains gravity, Newton wonders: Why are stars seen through telescopes fringed in rainbow colours?

Telescope uses lenses – glass discs whose thickness varies. Newton uses a simpler version: triangular wedge whose thickness varies (prism).

Newton places prism in the path of white sunlight coming through a slit in curtains at Woolsthorpe. Projected on darkened wall, he sees . . .

. . . fanned out into a 'spectrum', all colours of rainbow – red (r), orange (o), yellow (y), green (g), blue (b), indigo (i), violet (v).

(Not many people know this ... Roy. G. Biv is actually a character in *The Information*, the comic novel by British novelist Martin Amis.)

Newton places a second prism (the other way up) in the path of the spectrum and the colours miraculously recombine into white light.

Newton concludes, correctly, that white light coming from the Sun is in fact made of all the colours of the rainbow woven together.

What glass in a prism in fact does is bend different colours by different amounts, fanning out white light into its constituent colours.

Light is a wave (too small to see) & different colours are different sizes (wavelengths). Red light is about 2x wavelength of blue.

A rainbow is created by rain droplets, acting like myriad tiny prisms and splitting white sunlight into its constituent colours.

Back surface of droplet acts like tiny mirror. Light can re-emerge after 1 reflection or 2: so often 2 rainbows, 2nd with colours reversed.

The angle between the incoming and outgoing light ray is 42° (the 'rainbow angle'). For the secondary bow the angle is 51°.

A rainbow is in fact a circle. However, because the horizon gets in the way, we see only part of the circle – a semicircular bow.

Newton overcame problem of rainbow colours fringing stars by replacing lenses with mirrors. Invented 'reflecting' telescope. That's genius!

2. Why is the sky blue?

Since air is patently transparent, this is far from obvious!

Explanation of why sky is blue was found in late 19th century by English physicist Lord Rayleigh (winner of 1904 Nobel Prize for physics).

Key fact 1: light is a wave like a ripple spreading on a pond. This is far from apparent since size of wave (wavelength) too small to see.

Key fact 2: White sunlight, as Newton discovered, is made of all colours of the rainbow, from blue (smallest wavelength) to red (longest).

Key fact 3: Molecules of oxygen/nitrogen in the air just happen to be of a size that deflects (scatters) blue light far more than red.

Consequence: as white sunlight comes down through air, blue light is preferentially removed (scattered). Creates diffuse blue background.

As Sun sets on horizon, it turns red since light must travel through more atmosphere, subtracting 100% of the blue light, leaving only red.

If the particle size in atmosphere changes, so too can sky colour. Sky goes red if there are pollutants or dust from volcanic eruption.

If particle size is just right, you can even get a blue moon. One possible origin of the phrase 'once in a blue moon', meaning 'rarely'.

Sky on Mars can be pink or yellow since colour depends only on size of particles hoisted into the thin atmosphere by dust storms.

High up in Earth's atmosphere, there are few air molecules to scatter sunlight. The sky, instead of blue, is therefore inky black.

3. Why is the rising Full Moon so large?

Quick answer: it isn't. Yes, the Moon does indeed appear huge when low on the horizon (as does the Sun). But it's just an optical illusion.

Here's how to prove it. Hold a small coin at arm's length. Compare the relative size of the rising Moon to the size of the coin. Then . . .

. . . repeat the same thing with the Moon high in sky. You'll find that the Moon's size is exactly the same. It's called the Moon illusion.

The same is true for the rising or setting Sun. But, since you usually don't stare at Sun, you notice the effect principally with Full Moon.

BTW: same effect seen with the constellations. The Plough seen just above distant buildings looks much larger than it does high in the sky.

So what causes effect? No one knows. May be related to our perception of the sky (incorrectly) as slightly flattened, not 100% spherical.

The Moon illusion is much less impressive from the middle of an ocean. Seems to imply trees/buildings on horizon are important ingredient.

Most probable reason Moon appears large is because it's in the same field of view as distant objects whose true size we know all too well.

If the Moon illusion is trick of mind, it should be absent if you fool your mind into seeing familiar things like trees as unfamiliar . . .

Bend over and look at rising Moon between your legs. Everything is upside down & unfamiliar and, hey presto, the Moon illusion almost gone.

In reality, the Moon on horizon is slightly 'smaller' than Moon above your head. Think about it: it is further away by one Earth radius.

Also, not every Full Moon has same size in sky. This is because Moon's orbit is elliptical, causing the Earth–Moon distance to vary a lot.

Thing to remember: Moon is actually a very small object in the sky. Artists always make it bigger than it should be in paintings/drawings.

4. What causes the phases of the Moon?

Appearance of Moon constantly changing: thin crescent, half-lit, gibbous, full etc. Time for complete cycle approx. 29.5 days (lunar month).

Key fact: unlike Sun, Moon doesn't emit its own light. Instead, Moon is visible only when illuminated by Sun – when it reflects sunlight.

The 'phases' are caused by changing illumination by the Sun: sometimes a lot of the Moon's surface is illuminated, sometimes only a little.

Like Earth, Moon has bright day side (facing Sun) & dark night side. Always half lit; no permanent dark side of the Moon (sorry Pink Floyd).

When the Moon and Sun are in about the same direction, the Moon is lit from behind. From Earth, we see its dark side. This is a New Moon.

About a week later, Moon has completed first 25% of orbit (First Quarter). Sun now illuminates Moon from the west. We see half-lit Moon.

After another week, the Moon is in the opposite direction to Sun. From Earth, we see its illuminated side. This is a Full Moon.

Finally, after completing 75% of its orbit (Last Quarter), Moon is illuminated from the east. West-facing hemisphere of Moon is now dark.

Memory aid: First Quarter Moon visible only during first half of night; Last Quarter Moon can be seen only during last half of night.

The Full Moon is opposite the Sun, so it rises around sunset, and sets around sunrise. This means it can be seen all night long.

Average lunar cycle lasts 29 days, 12 hours, 44 minutes, 2.9 seconds. This lunar month is still the basis of Jewish/Islamic calendars.

Earth also goes through phases, as seen from Moon. During New Moon, astronaut on Moon would see Full Earth, and vice versa.

5. What is a lunar eclipse?

A lunar eclipse occurs when the Earth blots out sunlight falling on Moon. Impressive phenomenon, mainly because of spooky red colour.

For lunar eclipses the Earth must be between the Sun and the Moon. So they can take place only at the time of a Full Moon.

Moon's orbit slightly tilted from Earth's equator. Full Moon tends to pass slightly above or below Earth's 'shadow' so no lunar eclipse.

During total lunar eclipse, Full Moon first enters outer, 'penumbral' shadow of Earth: only part of sunlight is blocked. Moon looks 'murky'.

Then, Moon enters central, 'umbral' shadow. A small bite is taken out of Moon, which grows larger & larger. Eventually, Moon fully eclipsed.

Surprisingly, Moon doesn't disappear altogether even if no direct sunlight falling on surface. Instead, glows with dark orange-red colour.

The eclipsed Moon's red colour comes from sunlight passing through Earth's atmosphere. Air molecules 'scatter' some light into the shadow.

To understand: imagine you're on the Moon during total lunar eclipse. You are in Earth's shadow, so Sun is blocked by Earth and invisible.

But Earth's atmosphere glows as red ring around dark planet, just as evening sky glows red after sunset. Result: Moon gets faint red hue.

During lunar eclipse, Full Moon first gets dim, then dark & red. Many more stars become visible. Total phase may take up to 1 hour, 40 min.

Some lunar eclipses are only partial (part of Moon passes through umbral shadow), or just penumbral (almost invisible).

Next total lunar eclipses: 10 Dec 2011 (Asia, Australia), 15 Apr 2014 (Americas, Australia), 8 Oct 2014 (N. America, East Asia, Australia).

6. What is a total solar eclipse?

A total solar eclipse is without doubt the most spectacular natural phenomenon you will ever witness. Don't die without seeing one.

Solar eclipse occurs when the Moon passes in front of the Sun. Since Moon must be between Earth & Sun, can take place only during New Moon.

Not every New Moon produces a solar eclipse. In most cases, New Moon passes above or below Sun because Moon's orbit is slightly tilted.

During eclipse, lunar shadow crosses Earth, tracing narrow zone of 'totality'. To see totality, you need to be at right place, right time.

During partial phase, Moon covers ever larger part of Sun's disc. Eventually, temperature drops, light gets eerie, animals become alarmed.

In final minutes, shadow races over ground to observer, planets come out in day, last sunlight glows like gem on circular 'diamond ring'.

Then darkness sets in, bright stars appear. Moon is like black hole in the sky, surrounded by glow of Sun's outer atmosphere, or 'corona'.

Totality lasts only a few minutes. Very powerful emotional event (some people cry!). Spell is broken by first sunlight at end of totality.

Total solar eclipse is result of cosmic coincidence. Sun is 400 x Moon size, but also 400 x further away, so they appear same size in sky.

Sometimes Moon further from Earth than average. Smaller apparent size means can't cover all Sun. Rather than total, 'annular' solar eclipse.

Next total solar eclipses: 13 Nov 2012 (N. Australia, Pacific), 20 Mar 2015 (N. Atlantic, Svalbard), 9 Mar 2016 (Indonesia, Pacific).

7. Why are summers warm and winters cold?

Earth's orbit isn't a perfect circle. It's slightly squashed ('ellipse'), so distance from Sun varies. However, nothing to do with seasons!

If it was, every place on Earth would have same seasons. Instead, it's summer in northern hemisphere when winter in south; & vice versa.

Seasons in fact caused by tilt of Earth's rotational axis. Like a globe in a classroom, the Earth is tilted 23.5 degrees from vertical.

In June, Earth's northern hemisphere is tilted towards Sun; southern hemisphere away. Six months later (December) it's the other way round.

In summer, days are longer than nights. Also, Sun climbs higher in sky, heating ground more efficiently. Net result: higher temperatures.

In winter, nights are longer than days. Sun stays low above horizon and doesn't have enough strength to substantially warm Earth's surface.

In northern hemisphere, most sunlight on 21 June – midsummer's day (summer solstice); least on 21 Dec – midwinter's day (winter solstice).

Since the ocean and atmosphere respond slowly to varying sunlight, the warmest/coldest months are actually after summer/winter solstice.

Around 20 March & 22 September, Sun exactly above Earth's equator. Spring/autumn equinox. Day and night same length everywhere.

Every planet with an axial tilt has seasons. Martian seasons are like ours (similar tilt), but they last longer (longer orbital period).

However, varying distance from Sun plays greater role on Mars since it has much more elliptical orbit than Earth. Seasons more extreme.

8. What is a constellation?

Tens of thousands of years ago, people looked up at the night sky and imagined patterns in the randomly scattered stars.

Some star groupings appeared to resemble animals such as bulls, dogs, bears and snakes. Thus were born the constellations.

Later, other star groups were named after gods and figures of myth. The Roman polymath, Ptolemy (AD 90–168), listed 48 constellations.

Most famous include: Ursa Major (Plough), Orion, Leo (Lion), Cygnus (Swan), Taurus (Bull), Cassiopeia, Gemini (Twins), Hercules.

In late 16th century, Dutch sailors mapped southern skies & added new constellations like Tucana (Toucan) & Apus (Bird of Paradise).

Later, new inconspicuous constellations were added to the northern hemisphere such as Vulpecula (Little Fox) & Lacerta (Lizard).

Since 1930, there have been 88 recognised constellations. Every location in the night sky is assigned to one or other constellation.

The stars in a constellation can be at vastly different distances and are usually not related, so constellations are apparent groupings.

A nearby star & an ultra-distant galaxy can belong to the same constellation – as long as they are neighbours in the sky.

Even as seen from Earth, the constellations change appearance over time very slowly, thanks to the proper motion of stars in the sky.

Some constellations are always visible; others never (except to people on the equator). Most can only be seen during a particular season.

Incas and Aboriginals also recognised 'dark cloud constellations': dark dust clouds in the Milky Way that resemble animals, like the Jaguar.

9. What is the zodiac?

Sun, Moon and planets move against the background of fixed stars. In other words they travel from one constellation to another.

In Sun's case, background constellations can't be seen of course. Nevertheless, path of Sun can still be deduced from observations.

Turns out Sun, Moon and planets are not free to wander anywhere in the sky. They never, for instance, show up in the Plough, or Orion.

Instead, motions of Sun, Moon and planets are always confined to band of 12 constellations circling sky: the constellations of the zodiac.

Zodiac constellations are best known: Aries, Taurus, Gemini, Cancer, Leo, Virgo, Libra, Scorpio, Sagittarius, Capricorn, Aquarius, Pisces.

Though name 'zodiac' related to 'zoo', not all of its constellations are animals. Many are human. And Libra (Scales) isn't even alive.

The yearly path of the Sun through the zodiac (which reflects Earth's orbital motion) is actually a circle on the sky called the ecliptic.

Long ago, ecliptic was divided in 12 equal parts (zodiacal signs), more or less corresponding to constellations (which have unequal sizes).

Astrology (superstition) claims a person's character depends on position of Sun, Moon and planets in these zodiacal signs at time of birth.

In fact, due to slow wobble of Earth's axis, signs and constellations no longer match; shift is about one constellation in 2100 years.

Also, ecliptic (Sun's path) crosses non-zodiacal constellation of Ophiuchus (Serpent Bearer), which doesn't play role in astrology.

The zodiac contains some bright stars: Aldebaran (Taurus), Castor and Pollux (Gemini), Regulus (Leo), Spica (Virgo), Antares (Scorpio).

This means frequent beautiful 'conjunctions' of Moon and/or planets with one of these stars. Sometimes stars are even occulted by Moon.

10. What is the Milky Way?

Milky Way is faint band of light spanning the night sky. Can be seen only from dark places (outside cities), on clear, moonless nights.

According to Greek mythology, it was mother's milk spilled by goddess Hera when she breastfed Herakles. Romans called it Via Lactea.

In Norse mythology, ghostly Milky Way (Vintergatan, or Winter Street) was path along which dead souls travelled to Valhalla (afterlife).

Galileo Galilei (1564–1642) was first to study Milky Way with telescope. Surprised to find it consists of countless faint stars.

William Herschel (1738–1822) and Jacobus Kapteyn (1851–1922) tried to deduce extent and 3D shape of Milky Way by counting stars.

We now know the Milky Way is a giant, flattened disc of stars, with spiral arms. Sun is in outer regions of disc, close to central plane.

So why do we see Milky Way as band of light circling the sky? Analogy: living in suburb of giant city, where all buildings are transparent . . .

City is pretty flat, so most light you see during night is in (horizontal) band around you, with concentration in direction of city centre.

Looking up or down, you only see few lights (tall buildings, subway stations). Likewise, Milky Way is projection of flat disc of stars.

Size estimates of Herschel and Kapteyn much too small. Also, they thought Sun near centre of Milky Way. Hoodwinked by light-absorbing dust.

It's like being in city's suburb on very foggy night: you only see lights out to certain distance, and you seem to be at the centre.

Real size, spiral structure and dynamics of Milky Way could be measured only after advent of radio astronomy (1950s): not hampered by dust.

11. What are shooting stars?

Watch the night sky 15 minutes and you'll see something move across the stars. If it blinks and has a red light, it's probably an aircraft.

Bright orange/slow motion? Thai-style sky lantern, often in groups. Steady motion, visible for minutes? Artificial Earth-orbiting satellite.

As bright as planet Venus? Probably International Space Station (ISS). Follow @twisst for personalised, location-based alerts on Twitter.

But a star-like object that streaks across sky & is visible for just one or two seconds is almost certainly a meteor – 'shooting star'.

Not related to real stars at all. As meteor name implies (think meteorology), they arise high in Earth's atmosphere, at ~80 km up.

Their cause? Grain of sand/small pebble from space entering atmosphere (at 11 km/s or so). Heated to incandescence by friction with air.

The larger the pebble the brighter the meteor. The brightest ones are called fireballs. Can leave a faint trail lasting tens of seconds.

If large enough, burnt remnant can reach ground as a 'meteorite'. Hard to find unless falls on snow (Antarctica) or desert sand (Sahara).

Meteors often connected to comets, which lose dust as they orbit Sun. If Earth moves through dust, more meteors than usual: meteor shower.

Meteors in shower appear to originate in one region in sky: the radiant. Similar perspective effect occurs if you drive through snow storm.

Showers occur yearly around same date. Famous: Perseids, around 12/13 Aug. Named because radiant is in constellation Perseus.

Others: Quadrantids (4 Jan), Lyrids (22 Apr), Draconids (9 Oct), Orionids (22 Oct), Taurids (6 Nov), Leonids (17 Nov), Geminids (14 Dec).

12. How many stars can I see?

It depends. On a crystal-clear, moonless night, far from city lights, a few thousand stars are visible to the unaided eye.

From a big city, only the very brightest stars can be seen. Fainter ones are washed out by light pollution, a curse to (amateur) astronomy.

Greek astronomers ranked stars according to brightness (magnitude). Brightest stars: magnitude 1; faintest visible with naked eye: mag 6.

Magnitude scale still in use, but made more precise. Difference in mag of 5 corresponds to factor 100 in brightness (1 mag to factor 2.512).

Also, brightest stars turn out to be brighter than mag 1. And stellar brightness can be measured with precision of 0.01 mag.

Betelgeuse: mag 0.50, Vega: mag 0.03, Sirius (brightest star in the sky): mag -1.46. Negative numbers imply greater brightness (Sun: -26.8).

Only 50 stars are brighter than mag 2 (visible from city); 500 brighter than mag 4; 5000 brighter than mag 6 (limit naked-eye visibility).

Using a telescope greatly increases the number of stars you can see. Small amateur telescope reveals stars as faint as mag 10: 340,000.

Hubble Space Telescope has revealed stars of mag 30 – few billion times as faint as your eye can see.

'Apparent brightness' depends on distance. Betelgeuse appears fainter than nearby Sirius, though in reality it pumps out much more light.

'Absolute brightness' is measure of true luminosity of star. Defined as apparent brightness if object were at 10 parsecs (32.6 light years).

Absolute brightness of Betelgeuse is mag -5.1. Of Sirius: mag +1.43. So Betelgeuse is few hundred times more luminous than Sirius.

The Earth

13. How do we know the Earth is round?

Not obvious. Apart from wrinkles like mountains, the Earth seems flat. But this is because it's big and its curvature is imperceptible.

Abundant evidence round . . . Ships at sea disappear over horizon while still sizeable. On a flat Earth, would dwindle to dots first. And . . .

. . . During a lunar eclipse, when the Earth passes between the Sun and Moon, the Earth's shadow on the Moon is curved. And . . .

. . . If people sail or fly far enough in one direction, eventually they return to their starting point. And . . .

. . . Even mutual airline distances between any 4 cities – 1 to 2, 2 to 3 etc. – betray roundness. Would be different if Earth was flat. And . . .

. . . There are abundant photos of Earth taken from space – particularly from the Moon – which show our planet is quite definitely round!

In 240 BC, Eratosthenes, the chief librarian of the Museum at Alexandria, becomes the first person to estimate the size of the Earth.

At summer solstice, a vertical pillar in Syene (Aswan) has no shadow, since Sun is overhead. But a pillar at Alexandria does have a shadow.

It turns out the Sun is 7° from the vertical in Alexandria – about 1/50 of a circle. Distance between Syene and Alexandria is known. So . . .

. . . Eratosthenes calculates the Earth's circumference to be about 39,000 km, which is only 1000 km out.

Actually, Earth is not a perfect sphere. At equator, ground is rotating at about 1700 kilometres an hour. Planet's waistline bulges outward.

Unevenness in the planet's molten interior also makes the average level of the crust undulate, creating a knobbly figure called a 'geoid'.

14. Why are our feet glued to the ground?

In a word: gravity! Gravity is a universal force of attraction between all masses. As far as we know everything in the universe feels it.

There is a force of gravity between you and anyone who happens to be standing next to you; between you and the coins in your pocket.

But gravity is a relatively weak force. Hold your arm out. The cumulative gravity of all the matter of the Earth cannot pull it down.

Force of gravity, though weak, grows with mass. While negligible for small bodies, it's appreciable for big bodies – Earth, Sun, galaxy.

BTW, gravity is mutual. Pull of Earth on you = pull of you on Earth. Earth is less affected because it's bigger & it takes more to move it.

The man who wondered why he was attracted to big women but big women were not attracted to him was saying something profound about gravity!

The gravity of the Earth keeps our feet firmly glued to the ground and keeps the Moon trapped in perpetual orbit around the Earth.

From motion of Moon, Newton deduces gravity weakens with square of distance from mass. 2 x distance = 4 x weaker; 3 x distance = 9 x weaker.

Newton also proves that orbit of a planet under influence of 'inverse square law' of gravity is an ellipse, as observed by Johann Kepler.

Actually, Newton merely 'describes' behaviour of gravity. A better description is obtained by Einstein: general theory of relativity (1915).

According to Einstein's theory, matter (energy) tells space-time how to 'warp'; warped space-time (gravity) then tells matter how to move.

So Earth creates valley in space-time like depression made by bowling ball on trampoline. Other masses – like you – fall into it.

Neither Einstein nor Newton guessed what gravity *is*. Believed to be exchange of particles (gravitons), like balls between tennis players.

Problem: Despite heroic effort, none have found a description of gravity in terms of gravitons. 'Quantum' theory of gravity remains elusive.

15. What makes the Earth special?

Three reasons: life, life, life. Earth is only planet that boasts biology. But it also has other special properties, maybe related to life.

Of the four rocky planets of inner Solar System, Earth is the only one with water on its surface – important for origin & survival of life.

Venus and Mars, just after birth, probably also had water. On Venus (closer to Sun), oceans evaporated. Became ultimate 'greenhouse' planet.

Mars, being smaller than Earth, lost its heat more rapidly. Also lost most of its atmosphere/water vapour to space. Remaining water froze.

Earth has 'Goldilocks' combination of size and distance to Sun. If closer to Sun: too hot, like Venus. If much smaller: too cold, like Mars.

Earth also only rocky planet with large moon. Gravity of Moon righted Earth whenever axis tipped, so keeping climate stable for life.

Radioactivity kept Earth's core molten. Slow, charged flows generate magnetic field. Shields life against deadly particles from Sun/space.

Finally, Earth is only planet in Solar System with plate tectonics (see next page), which prevents build-up of CO_2 in atmosphere.

Earth may be giant 'self-regulating complex system' (Gaia hypothesis), with biology and geology conspiring to retain planet's habitability.

16. What are plate tectonics?

In 1620, Englishman Francis Bacon notices that the coastlines of Africa and South America are like jigsaw pieces that would fit together.

At the beginning of the 20th century, German Alfred Wegener wonders: could the continents have been joined at one time, then drifted apart?

Wegener dies in 1930 on a trip to Greenland. Tragically, he never sees the triumph of his hugely controversial idea – 'continental drift'.

By the late 20th century, Wegener's big idea is confirmed and fleshed out into the comprehensive modern theory of 'plate tectonics'.

Earth's skin (lithosphere) floats on molten magma. 2 types of crust: oceanic – thin/dense; continental – thick/light, which floats higher.

Lithosphere is fragmented into 'plates'. Where two continental plates collide, crust buckles upward, creating mountains like the Himalayas.

If continental & oceanic plates collide, oceanic dives under, buckling plate above. Result: mountains (Andes) & volcanoes (through friction).

When plates pull apart, at 'mid-ocean ridges', lava wells up & fills the gap, creating new crust. Ocean spreads. Atlantic was once a puddle.

Remarkably, in Africa today, we can see a new ocean being born. At Afar in Ethiopia, 3 plates are pulling apart. Gap will fill with water.

According to plate tectonics, 250 million years ago, Earth had a single 'supercontinent' (Pangaea). Its break-up created today's continents.

The driving force of plate tectonics is hot magma rising and cool magma sinking inside the Earth, much like water heated in a saucepan.

Heat source: radioactivity in rocks. Initially, this melted Earth, so dense iron sank to core & light rock (lithosphere) floated to surface.

No one knows what fractured lithosphere into plates in the first place. Could be simply cooling & shrinkage of Earth. Or impact from space.

Liquid water is crucial in lubricating motion of plates. Venus, roughly same size as Earth but with no water, does not have plate tectonics.

17. Why is the inside of the Earth molten?

It isn't. At least not in the planet's very centre. Earth has a solid inner core and a liquid outer core. Both consist of iron and nickel.

Normally, iron melts at 1536 °C. But a material's melting point rises at higher pressure. In inner core, pressure too high for iron to melt.

Solid inner core has diameter of 2430 km – 70% of size of Moon. Temperature: 5430 °C. Pressure: ~350 gigapascal (3.5 million atmosphere).

Molten outer core is about 2250-km-thick layer. Temperature ranges from 4400 to 6100 °C. Probably also contains sulphur and oxygen.

Current-carrying flows in outer core generate Earth's magnetic field. If whole core were solid, Earth wouldn't have magnetic field.

Earth is 'differentiated' body: heavy elements (iron, nickel) have sunk to centre because of gravity. 'Stratified' internal structure.

So why is interior of Earth so hot? Two reasons: relic heat of planet's formation and heat from decay of radioactive elements.

Earth formed from colliding and merging protoplanets. Generated lots of heat. Early Earth completely molten; made differentiation possible.

Radioactive elements (uranium, thorium, potassium) slowly decay into lighter elements. This also produces heat in Earth's core and mantle.

Earth slowly cools by losing internal heat to cold of space. But: large planet contains more heat (from both sources) than small planet.

Also, large body has smaller heat-losing area relative to volume. Cools more slowly. For same reason, adult cools more slowly than baby.

All geologic activity (volcanoes, earthquakes, mountain-building) is driven by heat flowing from interior of Earth to the surface.

18. How do we know the age of the Earth?

Question of the Earth's age is bound up with age of the Sun, since Sun cannot be younger than Earth (or the Earth would have frozen solid).

How long the Sun has been shining depends on how much heat it is pumping out – something measured in early 19th C – and its energy source.

In the 19th century, in a world powered by steam, it was obvious for physicists to wonder: is the Sun a giant lump of burning coal?

How long could Sun-sized lump of coal – the mother of all lumps of coal – maintain Sun's heat output before going out? About 5000 years.

5000 years was too little even for Irish Archbishop Ussher, who deduced from Bible Earth was created Sunday 23 October 4004 BC – at 9 a.m.

Geology and biology tell us that the Earth must be at least hundreds of millions of years old. Has taken ages for mountains to rise up . . .

. . . and for living things to evolve from a common ancestor. Physics also has something important to say about the age of the Earth . . .

Radioactive uranium in rocks decays into lead at a known rate, so ratio between two can be used as a 'clock'. Earth billions of years old.

In 1907, American physicist Bertram Boltwood radioactively dates rocks from Sri Lanka and finds they are a whopping 2.2 billion years old.

In fact, oldest rocks found on Earth are about 4 billion years old. But the Earth, of course, must be older still. The question is: how old?

Best age comes from radioactive dating of meteorites, builders' rubble left over from the birth of the Solar System: 4.55 billion years.

So the Sun and the Earth are about a third as old as the universe, which burst into being in Big Bang about 13.7 billion years ago.

Sun has been burning million x longer than it could if it was made of coal. So energy source must be million x as concentrated as coal.

An energy source million x as concentrated as coal exists: nuclear energy. Sun is 'fusing' hydrogen into helium. The byproduct is sunlight.

19. What protects us from the dangers of space?

Space is a dangerous place. It's a bitterly cold vacuum, pervaded by lethal radiation, deadly particles, meteorites and killer asteroids.

Earth is relatively safe from these cosmic dangers. We're protected by a thin layer of air (our atmosphere) and an invisible magnetic field.

Charged particles, like protons (hydrogen nuclei) and electrons, zip through space at near-lightspeed. High energy makes them dangerous.

They are produced by the Sun (solar wind/storms), and by supernovae and violent 'active' galaxies. Such particles are known as cosmic rays.

Astronauts on Moon or on trip to Mars could be killed by radiation from giant solar flare, or develop skin cancer from cosmic rays.

Earth's magnetic field deflects these particles, so most of them don't reach us. Moon and Mars (very little field) more dangerous places.

Most lethal high-energy radiation (UV and X-rays, mostly from Sun) is absorbed by air molecules. Otherwise, they too would cause cancer.

Small meteorites are slowed down, heated and vaporised by atmosphere. On airless Moon, small pebbles would puncture spacesuit/moonbase.

Still, there are many cosmic events that can endanger life on Earth. But most are very unlikely to occur during a human lifetime.

A truly giant solar flare could melt power stations and break down grid systems and communications networks, causing widespread chaos . . .

Or Earth could be hit by small asteroid or comet. A 1 km object would devastate a continent; a 10 km object could cause global catastrophe.

The good news (kind of): both radiation-induced genetic mutations & cosmic impacts drive evolution. Without them, we wouldn't be here.

20. What causes ice ages?

No one knows for sure. There's probably not one single cause. The long-term evolution of Earth's climate is not very well understood.

Geological evidence for ice ages was first discovered in early 19th century. Idea that ice ages recur was taken seriously only in 1870s.

Earth has experienced at least five major ice ages over past 2.5 billion years. Most last for tens or hundreds of millions of years.

Most severe was 850–630m years ago when ice covered almost all of the Earth. This 'Snowball Earth' was probably ended by massive volcanism.

Current (Pleistocene) ice age began 2.58m yrs ago. In past 740,000 years, 8 glacial periods separated interglacials. We're in interglacial.

Glacial periods first repeated every 41,000 years; then every 100,000 years. Current (Holocene) interglacial started 10–20,000 years ago.

Such long-term changes in temperature can be caused by changes in the amount of 'greenhouse' gases in the atmosphere that warm the planet.

Another possible cause is a change in position of continents, due to movement of 'plates'. May influence ocean currents and climate.

During 1st World War, Serbian civil engineer Milutin Milanković proposed that slow changes in Earth's orbit could cause ice ages.

Indeed, axial tilt of Earth varies from 22.1° to 24.5° every 41,000 years; the elongation of its orbit varies every 100,000 & 400,000 years.

Looks like 'Milankovitch cycles' play a role in creating glacial periods & interglacials. But the mechanism by which they do this not clear.

And changes in Earth's orbit and orientation are much too small to have caused major ice ages, let alone the Snowball Earth catastrophe.

The Moon

21. How big and how far away is the Moon?

The Moon is our nearest cosmic neighbour. It's also our only natural satellite, and the only other celestial body that humans have stood on.

Average Earth–Moon distance (centre–centre) is 384,400 km. If you could drive to Moon at 100 km/h non-stop, it would take almost 6 months.

Moon's orbit is not a perfect circle but an ellipse. Distance from Earth varies from 362,000 km (perigee) to 407,000 km (apogee).

At perigee, Moon looks a bit larger than average. If perigee coincides with Full Moon, it also looks much brighter.

Orbiting at about 3600 km/h, the Moon takes 27 days, 7 hours and 43.1 minutes to go once around the Earth.

Time between successive Full Moons (Moon opposite Sun in sky) is longer (29d, 12h, 44m), because Earth also moves around Sun in meantime.

The Moon, at 3476 km across, is 27.3% of Earth's diameter. It has 7.5% of Earth's surface area; and only 2% of its volume.

Moon has an iron core/rocky mantle. It is less dense than the Earth because it contains less iron and it is compressed less by weak gravity.

The Moon has 1/81 of Earth's mass. It has only 1/6 the Earth's surface gravity, which means you would weigh 1/6 as much as on Earth.

Moon is biggest satellite compared with its parent planet, and the 5th largest in the Solar System – after Ganymede, Titan, Callisto and Io.

22. Why doesn't the Moon fall down?

This is not a stupid question. After all, if you throw a ball up in the air, it always returns, pulled back down by gravity of the Earth.

Newton solves puzzle with this picture. Cannon fires ball, which loops through air & comes down. Bigger cannon fires ball faster/further.

Now imagine extremely big cannon. It fires a ball so fast and far that the curvature of the Earth becomes significant.

Just as fast as the ball falls, the Earth's surface curves away from it. So the ball never comes down! It . . .

. . . ends up in orbit, falling forever – in a circle. The speed necessary to achieve this is 27,400 km/s.

Likewise, Moon is perpetually falling towards Earth, never reaching it. But, at its distance, speed is not 27,400 km/s but only 3700 km/s.

Newton's genius is to realise that an apple dislodged from a tree and the Moon are both falling (from which he deduces law of gravity).

Artificial satellites are also falling while orbiting the Earth. But, if close, the drag of the atmosphere eventually brings them down.

Speed needed to reach Earth orbit is huge. But on a small asteroid with low gravity, you could run fast enough to go into orbit.

23. Is there a dark side of the Moon?

Yes, there is. The Moon gets its illumination from the Sun. So, at any time, it has a bright dayside & a dark nightside, just like Earth.

Common mistake. People call far side of the Moon, which is turned away from Earth, its dark side. But it's not always dark, of course.

At New Moon, Moon is more or less in direction of Sun. Although Earth-facing hemisphere is dark, far side of the moon is fully illuminated.

From Earth, we only ever see the near side of the Moon. Far side was unseen until photographed by Soviet *Luna 3* space probe in October 1959.

Seems as if Moon doesn't rotate on its axis. But it does. Rotation period is exactly as long as orbital period around Earth.

Many planetary moons have 'synchronous rotation'. Caused by 'tidal forces' of parent planet, braking rotation until one side faces parent.

You might think that only half of the Moon's surface is ever visible from the Earth. But, actually, it is more like 59% . . .

Moon's speed varies because of elliptical orbit. Rotation is constant, though. Result: from Earth, Moon appears to wobble ('librate').

Main difference between two faces of Moon: no large volcanic plains (lunar 'seas', or *maria*) on far side, probably because crust is thicker.

In distant future, Moon will 'brake' rotation of Earth so that it will always keep same face towards Moon, like Pluto and its moon Charon.

Last words on Pink Floyd album *Dark Side of the Moon*: 'There is no dark side of the moon really. As a matter of fact, it's all dark.' True.

24. Why is the Moon cratered?

The Solar System contains a lot of builders' rubble left over from its birth. The material is in the form of rocky asteroids & icy comets.

Over the 4.55-billion-year history of the Solar System, planets and moons have been in a shooting gallery, bombarded by debris from space.

Lunar craters are impact scars. With no weather/ground movement to erase them (as on Earth), they survive (unless blasted by later impacts).

The craters on the Moon are a history book in the sky. By learning to read it, we can also piece together the impact history of the Earth.

Biggest impacts occurred during the Late Heavy Bombardment, 3.8 billion years ago. Impacts were so huge they punctured the lunar crust.

Lava welled up to fill the 'impact basins', creating the dark lunar *maria* (plural of *mare*, which is Latin for 'sea').

LHB was caused when Jupiter & Saturn worked in concert to stir up the asteroid or comet belt, sending bodies as big as Los Angeles our way.

Some craters exhibit 'rays' from ejected debris, like 93-km Copernicus, formed when asteroid size of Key West struck ~800m years ago.

Though Earth's craters are mostly erased, some survive, like 1.2-km Meteor Crater, Arizona, made ~50,000 BC by objects size terrace houses.

Also, 180-km Chixculub crater, partly under sea off Yucatán. Believed created by 10-km asteroid that killed dinosaurs 65 million years ago.

25. How does the Moon influence the Earth?

Twice a day the sea advances up beaches and then retreats. Such 'tides', which were first explained by Isaac Newton, are caused by the Moon.

Contrary to popular belief, the Earth's tides are not caused by the gravity of the Moon but by *differences* in the gravity of the Moon.

Moon's gravity pulls strongest on the ocean facing it, less hard on the centre of the Earth, and least hard on the ocean facing away . . .

. . . So, oceans bulge in two directions – on one side because water pulled away from Earth; on other, because Earth pulled away from water.

As the Earth turns on its axis once every 24 hours, the two tidal bulges travel around the oceans, at each point creating two tides a day.

Actually, the Moon's gravity pulls on the tidal bulges. This acts to 'brake' Earth's rotation. The Moon reacts by retreating from the Earth.

The Moon pulls tides in rock as well as water, though smaller since rock is stiffer. Such tidal stretching may help to trigger earthquakes.

Large Hadron Collider near Geneva observed to expand & shrink twice a day as Moon stretches & squeezes 27-km ring of 'atom smasher'.

The Sun also creates tides in oceans, though 1/3 as big as Moon. When the Sun and the Moon are pulling together, we get the highest tides.

High tides, winds & a funnel-shaped river can create tidal 'bore' – hump of water that keeps shape for many kilometres & can even be surfed.

In the past, when the Moon was closer, tides were higher than today. At its birth, when the Moon was 10 times nearer, 1000 times higher.

As well as tides, Moon can blot out Sun. 'Total eclipses' terrified ancients. Banged pans to scare away monster eating Sun. (Always worked!)

Total eclipses changed history. During battle Lydians v Medes, Turkey, 585 BC, land into plunged darkness. Bad omen. Armies laid down arms.

26. What if we did not have the Moon?

Very probably, we wouldn't be here!

Key fact 1: Moon unusually big – far bigger than any other satellite compared with parent planet. Earth-Moon essentially a 'double planet'.

Key fact 2: Life on Earth could not have evolved without a stable climate over billions of years. Our big Moon stabilises the climate.

It works this way: If Earth tips over – as spinning tops tend to do – sunlight reaching ground varies. Causes catastrophic climate change . . .

But, if the Earth tips over, gravity of big Moon pulls it upright again (Mars, with no big moon, experiences catastrophic climate changes).

Big Moon also drove colonisation of land. By pulling large tides, it left ocean margins high & dry. Fish, stranded, evolved lungs & walked.

A big Moon has also been crucial for science. By blotting out the Sun (total eclipse), it makes stars visible close to the solar disc . . .

. . . In 1919, such stars revealed the bending of starlight by the Sun's gravity, confirming a key prediction of Einstein's theory of gravity.

In 1972, in 'The Tragedy of the Moon', Isaac Asimov claimed that, if Venus had moon not Earth, science would have arisen 1000 years earlier.

Why? If Venus was orbited by a visible moon in the night sky, the Church-supported Earth-centred cosmos would have been an untenable idea.

27. How many people have been to the Moon?

Just twelve people have walked the Moon. Only nine are still alive. The youngest is Charles Duke (*Apollo 16*), born 3 October 1935.

President John F. Kennedy announced Apollo Moon programme in famous speech to US Congress, 25 May 1961: '. . . before this decade is out . . .'

Apollo 8 and *10* flew to Moon and back again without landing. In 1970, *Apollo 13* also had to return without landing, because of an explosion.

Apollo 11, *12*, *14*, *15*, *16* and *17* landed on Moon. In each case: two astronauts walked on surface; one stayed in command module orbiting Moon.

21 July 1969. *Apollo 11* lands. 1st man on Moon is Neil Armstrong (38); 2nd Buzz Aldrin (39). They spend 2 hours, 24 minutes on surface.

Lunar Roving Vehicle (moon buggy) enabled greater distances to be covered by crews of *Apollo 15* (27.8 km), *16* (26.6 km) and *17* (35.9 km).

14 December 1972. Last man on Moon is Gene Cernan (38) of *Apollo 17*. Plans for *Apollo 18*, *19*, *20* cancelled for lack of political support.

Apollo astronauts returned 382 kg of Moon rocks. Detailed analysis revealed that the Moon was probably ripped from the newborn Earth.

Three astronauts flew to Moon twice: Jim Lovell (*Apollo 8/13*), John Young (*Apollo 10/16*) & Gene Cernan (*Apollo 10/17*). Lovell never landed.

Number of people who flew to Moon (with or without landing): 21. No other astronauts have ever been more than a few hundred km from Earth.

28. Will footprints last forever on the Moon?

No. But they will last a very long time!

Moon has no wind or rain to erase footprints left by *Apollo* astronauts. On the other hand, it has a 'rain' of micrometeorites from space.

Micrometeorites, often smaller sand grains, create 'shooting stars' when burn up Earth's atmosphere. Invisible rain on Moon since no air.

Bombardment of Moon by micrometeorites over billions of years has shattered surface rock into thick layer of fine dust called 'regolith'.

Once upon a time there was a real fear that parts of the Moon were covered in a deep layer of dust and spacecraft would sink without trace.

In Arthur C. Clarke's 1961 novel, *A Fall of Moondust*, the lunar dustcruiser *Selene* sinks with all its passengers in sea of lunar dust.

The continual bombardment by micrometeorites turns over the top centimetre of lunar 'soil' about every 10 million years (lunar gardening).

The footprints left by astronauts will therefore not last forever. Nevertheless, they have good chance of outlasting human civilisation.

Lunar dust grains are very different from smooth sand grains on beach. Micro-impacts shatter rock into grains like tiny melted snowflakes.

Particles of moondust are like burrs. Astronauts could not get them off spacesuits. Got into every nook & cranny. Said smelt of gunpowder!

29. Is there water on the Moon?

The large dark patches on the Moon were once thought to be seas (*maria* in Latin). However, we now know that they are volcanic lava plains.

Surface water is impossible on the Moon. Without an atmosphere it would immediately boil away into space. The Moon is therefore bone dry.

Analysis of *Apollo* Moon rocks seemed to confirm idea of dry Moon. Tiny amounts of water found thought to be contamination from astronauts.

But, in 2009, Indian spacecraft *Chandrayaan-1* detected the 'spectral' signature of water (H_2O) or hydroxyl (OH) on lunar surface.

Observation was confirmed by other spacecraft: *Cassini* (en route to Saturn) and *Deep Impact* (passing Earth/Moon on way to Comet Hartley).

Small amounts of water: just 0.1% (1 litre per tonne). Probably formed by solar wind (hydrogen nuclei) combining with oxygen-rich minerals.

Water molecules are only loosely bound to lunar rocks. Means that water slowly creeps away from lunar equator to colder polar areas.

Lunar water accumulates as ice in deep craters at lunar poles. Their floors, which are in permanent shadow, never feel warmth of the Sun.

On 9 October 2009, engineers crashed *LCROSS* spacecraft into polar crater Cabeus. At least 100 kg water detected in plume rising from impact.

Water on Moon is crucial for establishment of any future moonbase. Essential not only for drinking but for making rocket fuel too.

However, according to LCROSS, lunar water exists not as large sheets of ice but intermingled with lunar soil, making it harder to extract.

30. Is the Moon a dead world?

Probably you think the Moon is stone-cold dead, its crater-strewn desolation untouched by the hand of change. But think again . . .

Since well before the invention of the telescope, at a rate of once every few months, strange lights have been reported on the Moon.

For instance, on 18 June 1178, five monks at Canterbury Cathedral in south-eastern England reported seeing an explosion on the Moon.

Mysterious lights on the Moon – dubbed Transient Lunar Phenomena, or TLPs – are one of the Moon's greatest enduring mysteries.

TLPs are localised events seen with telescope. They cover more than a square kilometre and can last anything from a minute to a few hours.

TLPs involve the brightening, dimming or blurring of the surface. Sometimes, there is also late-time colour change to a ruby-red glow.

Intriguingly, most TLPs occur at 6 locations on the Moon, notably in the craters Aristarchus (450 km diameter) and Plato (100 km).

All 6 sites are associated with outgassing of radon-222 as detected by *Apollo 15* & *16*, & *Lunar Prospector*, which orbited Moon in 1998.

Common feature 6 sites? Big craters from relatively recent impacts. Or at boundaries *mare* basins, created by huge impacts 3.8 bn years ago.

Sites are also locations of the several 100 moonquakes recorded by seismometers left on Moon by all but 1 of the 6 *Apollo* missions.

Crucially important feature of the 6 sites: cracks in the crust from ancient impacts, which allow sub-surface gas to escape into space.

Seems gas working its way out from lunar interior builds up pressure before bursting plug of lunar soil & exploding into space as TLP.

A mere 1/2 tonne of gas escaping into space could create cloud a few km across, lasting 5–10 min. Could change reflectivity of surface.

Origin of gas? 'Tidal' squeezing of Moon by Earth grinds up 1 aircraft carrier's mass of rock per year. Releases ~100 tonnes of gas/year.

A TLP happening at a human landing site would be very dangerous. *Apollo 18* – cancelled before launch – was once scheduled for Aristarchus!

31. When will people return to the Moon?

NASA's *Apollo* programme came to an end four decades ago. Although unmanned spacecraft have visited the Moon since, no humans have.

Unmanned craft have big advantages: can visit Moon for longer, cover larger areas, collect more data – and are hugely cheaper.

Still, Moon is nearest neighbour in space. Practice for human spaceflight beyond Earth orbit. Stepping stone for visits to Mars and beyond.

In 2004, US president George Bush announced Constellation programme: return to Moon (2020) with newly developed spacecraft. Then on to Mars.

Heavy-lift *Ares* rockets would launch *Apollo*-like *Orion* spacecraft and *Altair* lunar lander. *Ares* rocket was successfully tested in Oct 2009.

But in 2010, President Obama cancelled Constellation, which was behind schedule & way over budget. New plan: manned flight to asteroid.

So no NASA plans at present for return to Moon. Also, European ESA no longer pursues manned flights to the Moon, focusing on Mars instead.

China, though, appears to be planning human flight to Moon. No official announcements yet, but it could be around 2024. Same true for Japan.

In 2024, it will be 52 years since humans last walked on Moon – almost as long as the interval between first powered aeroplane and sputnik.

As for human flights to Mars, space agencies have been promising this will happen in 30 years' time ever since the 1960s. Umm.

32. Where did the Moon come from?

The origin of the Moon has been a long-standing mystery. No other moon is anywhere near as big compared with its parent planet.

Apollo missions found key evidence. Moon made of material similar to Earth's mantle. Lunar rocks contain far less water than rocks on Earth.

1975. Big Splash theory, proposed by William Hartmann & colleagues. Shortly after birth, Earth was hit by Mars-mass body (dubbed Theia).

Because it was heavy, the iron core of Theia sank to centre of the Earth. Molten mantle splashed into space. Formed a ring around Earth.

The ring of cooling debris coalesced into the Moon – 10 times closer than today, 10 times bigger in sky, pulling 1000-times-bigger tides.

Theory explains why Moon is like Earth's mantle; why it has no iron core; why it is so dry (water driven off in super-hot, violent impact).

Apart from big Moon, smaller moons also may have formed, which later collided with larger one. Could explain thick crust on Moon's far side.

Problem: for the Earth not to have been destroyed by Theia, it must have slammed into our planet at an inexplicably slow speed.

Explanation of Richard Gott and Ed Belbruno: Theia actually shared Earth's orbit, the way 'Trojan' asteroids today share orbit of Jupiter.

So, believe it or not, long ago the Earth may have had a brother, burning brightly in the night sky!

If Theia formed at a stable 'Lagrange point', 60° ahead or behind Earth in its orbit, & was knocked out, it would have approached slowly.

Energy lost in pulling tides gradually sapped Moon of orbital energy. Moved out to present location. Today, Moon still receding at 4 cm/yr.

Lesson for ET life. Big Moon has kept Earth's climate stable for life. Yet creation requires rare event. Therefore ET life may be rare.

Space

33. What is it like in space?

In space no one can hear you scream. That's because sound is a vibration of the air, and in space there is no air to vibrate.

In space laser beams are invisible (sorry, *Star Wars* fans). That's because dust in air 'scatters' light from a laser beam into your eye.

In space it's mind-cringingly cold. That's because there are very few atoms to strike you and, in doing so, transmit any heat.

In fact, interplanetary space contains about 10 atoms per cm^3 (a good terrestrial vacuum 100,000 & air at sea level 30 billion billion).

Actually, with no air atoms to carry away excess heat, overheating is as serious as freezing. Spacesuits must be heated AND cooled.

In space, with no air to breathe, astronauts must carry their own air supply with them, normally in scuba-like bottles on their back.

In space there is no pressure. We live with 50-km column of air weighing down on us (2 elephants!). Astronaut suits must be pressurised.

If astronauts did not look like Michelin Man, nitrogen would bubble out of their blood (nitrogen narcosis – 'the bends') and kill them.

In space you are weightless. In orbit, constantly falling towards Earth (but never getting there!). In 'freefall', you don't feel gravity.

In space there is constant danger of radiation from Sun and beyond (cosmic rays). Terrestrial magnetic field provides umbrella on Earth.

Astronauts often report seeing peculiar flashes of light. Thought to be high-speed subatomic particles smashing through fluid of eyeball.

Space radiation hazard has been called the only showstopper for manned exploration of space. Trip to Mars exposes astronauts for 6 months.

Get the feeling humans weren't made for space?!

34. How does a rocket work in space when there is nothing to push against?

Key fact: According to Newton's third law of motion, for every action there is an equal and opposite reaction.

Definitely true when you run. Your feet push backwards against the ground (action) and the ground pushes you forward (reaction).

But, contrary to expectations, to get such a reaction, it is not necessary to push against anything external.

Imagine you are stuck on a sledge in the middle of a perfectly slippery (frictionless) ice rink. How can you get to the side?

Say, sledge is (conveniently) stacked with bricks. Throw them, 1 at a time. As you push them away, they push back (reaction). Sledge moves.

This is the principle of the rocket. Gases are expelled from back at high speed. Rocket reacts by moving forward. Action & reaction. Simple.

As a rocket expels 'reaction mass', it becomes lighter, so its exhaust becomes ever more effective at driving the rocket forward.

This effect, encapsulated in the 'rocket equation', was first described by deaf Russian schoolteacher, Konstantin Tsiolkovsky, in 1903.

Problem: Even the most powerful rocket fuel has insufficient oomph to lift its own weight plus the weight of a rocket into Earth orbit.

Tsiolkovsky's solution: multi-stage rocket. Throw away part of rocket when high up. Rocket lighter & easier to boost to orbit.

Using throwaway rockets is like driving to town & arriving home with only 4 tyres & steering wheel, & having to re-build car for next trip.

NASA threw away much of the Space Shuttle and had to re-build it for each trip. One reason why it cost about $0.5 billion to launch.

Key to efficient rocket – at least in space – is high exhaust speed, so less fuel mass is needed. Current 'chemical' fuels are inefficient.

Ultimate rocket would use annihilation of matter/antimatter. Provides biggest push for given mass, so fuel weight to carry would be minimal.

The Sun

35. Does the Sun have a surface?

The Sun is a giant glowing ball of gas so it does not have a solid surface like the Earth. But it certainly appears to have one. Why?

The Sun's 'surface', or photosphere, is where sunlight, having struggled very hard to get out of the solar interior, begins to find it easy.

Picture a crowded street. Progress is slow. You must zigzag around obstacles (people). Can't go straight. Same for light emerging from Sun.

Photon (light particle) emerging from core of Sun travels only 1 cm before it is scattered into another direction by an obstacle (electron).

If travelling in straight line, photon would take only 2 secs to get from core to surface. But zigzag path so tortuous it takes 30,000 yrs!

Today's sunlight is therefore about 30,000 years old. It was made at the height of the last ice age.

If sunlight-generating nuclear fires went out 29,000 years ago, we would not know for another 1000 yrs.

Actually, not quite true. Sun would take a million years to lose its stored heat. Its big heat capacity would mean we'd be safe for a while.

Eventually, after 30,000 years, photons emerge from the Sun's surface and fly in a straight line to the Earth, at 300,000 km/s.

At this speed, it takes only 8.3 minutes to cover 150m km to Earth (so, if the Sun suddenly vanished, we would not know for 8.3 minutes).

Sun's photosphere is defined as place where photons go from travelling in zigzag path to straight line, or 'stop walking and start flying'.

Although the photosphere is not a solid surface, it is sharp enough that the Sun looks like a disc (if viewed through a safe filter).

36. Why is the Sun hot?

Sun is hot for a simple reason. It contains a lot of mass. The huge amount crushing down on the core under gravity squeezes matter there.

When gas is squeezed, it gets hot, as anyone who's squeezed air in a bicycle pump knows. In Sun's core gas is crushed to ~15 million °C.

At such high temperature, matter dissolves into anonymous state – 'plasma'. Behaves same no matter what the matter (no pun intended).

Sun is billion billion billion tonnes mostly hydrogen gas. But a billion billion billion tonnes bananas in one place would be equally hot.

Key fact: Sun's temperature depends on AMOUNT of matter it contains, not its composition (though has small effect on bottling heat inside).

But matter squeezing down on core explains only why Sun is hot at this instant, not why it stays hot. That's a different question entirely.

Sun is constantly losing heat into space yet never cools. Therefore something must be replacing the heat as fast as it is lost. But what?

Answer: nuclear energy. Sun is 'fusing' cores, or 'nuclei', of lightest element, hydrogen, into next lightest, helium. Byproduct: sunlight.

Just about most inefficient nuclear reaction imaginable. On average, it takes 2 hydrogen nuclei in Sun 10 billion years to meet and fuse.

Be thankful. Because solar fusion is so slow, Sun will take 10 bn years to burn its fuel – enough for evolution intelligent life like us.

To get an idea of inefficiency of Sun, imagine your stomach & volume of core of Sun size of your stomach. Your stomach generates more heat.

Q: If solar heat generation is so inefficient, why is Sun hot? A: Because it has an awful lot of stomach-sized chunks stacked together!

37. What's it like inside the Sun?

Sun is great ball of gas, 1.4m km across. Mainly hydrogen (75%) & helium (24%). Towards centre, density/temperature increase enormously.

No neutral atoms. Atomic nuclei (positive charge) stripped of electrons (negative). This gas of charged particles is called a plasma.

Temperature in Sun's core: 15.7m °C; density: 160 x greater than water. Hot/dense enough to trigger sunlight-generating nuclear fusion.

Diameter of core is 350,000 km (25% of Sun; 27 x Earth). Within this region, 99% of Sun's energy is produced (275 watts/m^3 in very centre).

Surrounding core is 'radiation zone', 315,000 km thick. Temperature drops from 7 to 2m °C. Energy floods outwards via radiation (light).

Core and radiation zone rotate like solid bodies – turning at same rate throughout. However, the core probably rotates slightly faster.

Outer region of Sun (210,000 km thick) known as convection zone. Like boiling saucepan. Hot plasma rises, emits energy. Cool plasma sinks.

Rotation rate of convection zone varies with depth and latitude (faster at equator, slower at poles). Known as differential rotation.

Also, subsurface meridional flows – 'rivers of fire' – carry plasma from equator to poles (close to surface) and back again (greater depth).

Magnetic fields are carried along, wrapped up, stretched and twisted by moving plasma. Such pent-up magnetic fields drive solar flares etc.

38. What are sunspots?

Sunspots are short-lived dark splotches on bright face of Sun. Though hot, they look dark because they're much cooler than surroundings.

The biggest sunspots can be 80,000 km across – more than 6 times wider than Earth. They often appear in groups and can last for weeks.

Largest sunspots are visible with naked eye when Sun dimmed at sunset/sunrise. Reported by Chinese astronomers and European monks.

In June 1611, German amateur Johannes Fabricius was first to describe sunspots seen through a telescope. Slightly earlier than Galileo.

By observing sunspots, it was possible to determine that Sun turned about once every 25 days. But true nature of spots remained a mystery.

Temperature of sunspots is 3000–4000 °C, compared with 5500 °C of solar surface. Sunspots are about 1000 km deeper than their surroundings.

A sunspot is caused by localised strong magnetic field, which stops convection. Without hot plasma rising from below, surface cools & sinks.

Larger sunspots have a dark, central area, called the umbra, and a lighter, surrounding area, called the penumbra.

Number of sunspots (s) & groups (g) – expressed as Wolf number (10g + s, named after solar observer) – is seen as measure of solar activity.

Often around sunspot groups (active regions) there appear bright 'faculae', solar flares and other explosive events. All magnetic phenomena.

Other rotating stars – stars like Sun & especially red dwarfs – show periodic changes in brightness. These betray existence of 'starspots'.

39. What is the solar cycle?

Heinrich Schwabe observed Sun constantly, looking for hypothetical planet within orbit of Mercury. Hoped to catch speck crossing solar face.

Instead, Schwabe found slow variation in sunspot numbers. Spots every day 1828/1829. But 139 spotless days 1833. Same pattern next decade.

In 1843, Schwabe published theory that sunspots exhibit 10-year cycle. Theory confirmed by older observations back to Galileo's time.

True period of solar cycle later found to be closer to 11 years. From 1755 onwards, cycles have been numbered. Our current cycle is no. 24.

New cycle begins with few small spots forming at high latitudes. Later, closer to equator, more active regions erupt, with spots and flares.

During 'solar maximum', Sun as a whole emits slightly more energy (despite larger number of dark spots), especially in form of UV & X-rays.

Origin solar cycle probably related to subsurface flows of magnetised plasma and regular build-up of magnetic energy. But details a mystery.

In 1893, Scottish astronomer Edward Maunder discovered that from 1645 to 1710 solar activity was unusually low. Christened Maunder Minimum.

Cause of Maunder Minimum is unknown. Tree ring analysis reveals similar long minimum between 1420 and 1550. Could happen again any time.

During 20th century, Sun was unusually active, with strong maxima. However, solar minimum of last cycle, no. 23, was very deep and long.

Solar activity of current cycle 24 predicted to peak in summer of 2013. Given late & slow rise of new cycle, maximum will probably be weak.

40. What is the solar wind?

Sun blows charged particles into space, mainly protons (+) and electrons (–). This million-mile-an-hour hurricane is called the solar wind.

Solar wind first suggested by Richard Carrington (1859). Theory published in 1958 by Eugene Parker. Confirmed by Soviet craft *Luna 1* (1959).

Because of solar wind, Sun is losing mass: 1.8 million tons/sec, or 1 Earth mass in 150 million years. Tiny amount compared to mass of Sun.

Solar wind originates in solar corona (Latin for 'crown'): extremely hot (1–3 million °C) tenuous outer 'atmosphere' of Sun.

Corona is a million times fainter than surface of Sun. Visible only with special instrument (coronagraph) or during total solar eclipse.

High temperature of corona probably due to shock waves; exact mechanism unknown. Result: particles move so fast they escape Sun's gravity.

Wind has 2 components. Slow solar wind (400 km/s, 1.5m °C, from corona) and fast solar wind (750 km/s, 0.8m °C, from Sun's surface).

Most of fast solar wind is accelerated by magnetic energy through 'coronal holes': regions where magnetic field lines open up into space.

Also, explosions on Sun produce billion-tonne 'coronal mass ejections', or CMEs: huge plasma clouds, blown into space as solar storms.

Close to Earth, solar wind and solar storms buffet Earth's magnetic field. Create spectacular auroras and may even destroy power grids.

Solar wind blows bubble, ~3 bn km across, into interstellar space. This 'heliosphere' is region where Sun's magnetic field dominates.

Edge of heliosphere/transition into interstellar space has now been reached by *Voyager* spacecraft, launched in 1977 and heading to stars.

41. How dangerous are solar flares?

A powerful solar flare could destroy electrical infrastructure, returning us to the steam age. Fortunately, such super-flares are very rare.

First solar flare ever observed, and most powerful on record: 1 Sep 1859 (Richard Carrington, London). Electrocuted telegraph operators!

Solar flares are energetic explosions on Sun's surface, powered by magnetic energy. They are more frequent at solar maximum.

The total energy released in a powerful solar flare may be a million times the annual electricity consumption of the world.

Solar flares produce energetic X-rays. May knock out spacecraft electronics. And harm astronauts orbiting beyond protection of atmosphere.

Solar flares decrease lifetime of artificial satellites, by heating up/expanding upper atmosphere, so creating drag on orbiting bodies.

Astronauts outside Earth's protective magnetic shield also at risk from burst of energetic protons, travelling near-lightspeed, from flare.

Flares are often accompanied by slower coronal mass ejections, though nobody knows why. CME particles may reach Earth a few days later.

Charged particles penetrating Earth's magnetic field may cause 'geomagnetic storms'. Visual manifestations are stunning polar 'aurorae'.

Much worse: flares disrupt GPS signals and radio communication; currents induced in wires may knock out power grids and computer networks.

Technology vulnerable: a flare like that of 1859, if pointing our way, could plunge planet into night-time darkness for weeks/months.

Power loss would disrupt communication, fuel & food supplies, health care, global economy. Would be huge death toll from famines/epidemics.

Remedy? Flare alert system out in space. Power grids and communication networks could then be switched off intentionally to protect them.

42. Is Earth's climate affected by the Sun?

Earth's climate is powered by the Sun's energy. Tiny change in solar output would have dramatic effects on weather and climate.

Over its lifetime, the Sun, like all stars, has become steadily brighter and hotter. In distant future, Earth will become too hot for life.

On shorter timescales, effect is less clear-cut. However, measurements show Sun produces slightly more energy during solar maximum (0.1%).

Response of Earth's climate to these variations is probably too slow to be of any importance, as solar cycle lasts only 11 years on average.

But, the Maunder Minimum, period of low solar activity between 1645 & 1715, coincided with Little Ice Age. Europe 1 °C colder than normal.

Also, higher-than-average solar activity in 20th century may have contributed to global warming. But there is much controversy about this.

Apart from direct effect of the Sun (irradiance), low solar activity may cool planet by encouraging formation of sunlight-reflecting clouds.

Theory: during solar minimum, solar wind is less powerful, so more energetic cosmic rays from outer space can enter Earth's atmosphere.

Cosmic rays rip electrons off atoms in air; the resulting charged ions act as tiny 'seeds' around which water droplets (of clouds) condense.

Scientific consensus: global warming largely due to burning of fossil fuels. Sun may play secondary role, but precise contribution unclear.

Long periods of high/low solar activity almost certainly influence climate, so new Maunder Minimum would for a while offset global warming.

In the very long run, Earth will experience a runaway greenhouse effect, just like Venus. Oceans will boil away; planet will roast.

43. Will the Sun last forever?

Nothing, as they say, lasts forever. And that is just as true of the Sun as of everything else in the world.

Every second Sun turns about 400 million tonnes of hydrogen into helium, byproduct being sunlight. One day hydrogen supply will run out.

As hydrogen 'burns' to helium, helium 'ash' falls to centre of Sun. In about 5 billion years, no hydrogen will be left in the solar core.

The Sun, at 4.55 billion years old, is therefore about halfway through its steady hydrogen-burning, or 'main-sequence', life.

Accumulating helium ash makes solar core denser and hotter since helium is heavier than hydrogen. So Sun gets hotter as it loses heat!

Sun already 30% brighter than when born. (Q: So, why did newborn Earth not freeze solid? A: Greenhouse gases may have kept planet warm.)

In future, Sun's core will continue to get denser/hotter. Extra heat flooding out will puff up envelope to huge size, creating 'red giant'.

A red giant is cool like an ember but, by virtue of its enormous surface area, may radiate 10,000 times as much heat as the Sun.

Earth will be fried, turned into a blackened cinder. But will it actually be swallowed by the ballooning Sun? This is not clear.

Red giants puff off outer layers into space. So the Sun will lose mass & slacken its gravitational grip. Earth will move away.

So, although Sun will likely swell to encompass the Earth's orbit, by the time it does so, the Earth will no longer be there!

During red-giant phase, Sun will be profligate with its heat. Eventually, it will cool and shrink down to a super-dense 'white dwarf'.

A white dwarf is about size Earth (sugar cube of its matter heavy family car). It's a stellar ember, cooling & fading into invisibility.

The Sun, in the words of poet T. S. Eliot, will end 'not with a bang but a whimper'.

The Solar System

44. Where did the Solar System come from?

In the beginning was a cold (-260 °C), dark interstellar cloud of gas and dust, an inky blot against the background stars.

The cloud may have hung there forever, doing nothing, if not for a kick, perhaps from the blast wave of an exploding star (supernova).

About 4.55 billion years ago, the cloud began shrinking under self-gravity, its gas becoming squeezed together ever more tightly.

When a gas is squeezed, it gets hot. The outward force exerted by warming gas should therefore have stopped the shrinking gas in its tracks.

But molecular hydrogen, carbon monoxide etc shed heat as light (microwaves), which escaped cloud, robbing it of ability to oppose gravity.

Initially, the cloud spun slowly (since Milky Way is spinning slowly). But, as it shrank, it spun faster, like ice skater pulling in arms.

Cloud shrank quicker between poles than around waist, where outward 'centrifugal' force opposed gravity. Became a flat, spinning pancake.

At the centre of the cloud, gas squeezed & heated to millions of degrees. Sunlight-generating nuclear reactions switched on. Sun was born.

In the debris disc swirling around newborn Sun, dust grains hit & stuck, building ever bigger particles up to km-sized 'planetesimals'.

In last, violent stages of Solar System's birth, planetesimals repeatedly collided, gradually building up the planets, including Earth.

Simulations often show 10 Earth-mass bodies forming. Encounters with embryonic giant planets flung Earth's brothers into interstellar space.

But the Solar System was not born alone. Elsewhere in the vast stellar nursery of the cloud other stars and planets were also born.

Supernova explosions of nearby massive (fast-living) stars rocked the young Solar System. Nuclear supernova debris is found in meteorites.

45. What is a planet?

The word 'planet' comes from Greek *planētēs*, which means 'wanderer'. Planets are celestial bodies that move against the background stars.

In antiquity, seven planets were known: the Sun, the Moon, Mercury, Venus, Mars, Jupiter and Saturn. Earth was not considered a planet.

In 'heliocentric' world view of Nicolaus Copernicus (1543), planets are objects orbiting Sun: Mercury, Venus, Earth, Mars, Jupiter & Saturn.

Difference between star & planet: in general, star big & hot, emits light and heat; planet small & cold, receives light & heat from star.

Difference in night sky: planets usually appear brighter, don't twinkle as much as stars, and slowly change position between stars.

But not everything that orbits the Sun is considered a planet. Notable exceptions are comets and asteroids, also known as minor planets.

With discovery of many Pluto-like icy bodies in the 'Kuiper Belt', beyond orbit of Neptune, necessary to refine definition of 'planet'.

Three characteristics of object. 1) Must orbit Sun. 2) Must be spherical due to self-gravity. 3) Must have 'cleared its orbit' of debris.

Only eight bodies in Solar System satisfy these criteria: Mercury, Venus, Earth, Mars, Jupiter, Saturn, Uranus and Neptune.

A few bodies satisfy criteria 1 & 2, but not 3. Such objects, most notably asteroid Ceres & Kuiper Belt Object Pluto, dubbed dwarf planets.

Criterion 4 might be: planets must be less than ~14 Jupiter masses. More massive objects are brown dwarfs, exhibiting some nuclear fusion.

Some astronomers have suggested criterion 5: planet must form by aggregation of matter in debris disc swirling around a newborn star.

While only eight planets are known in our Solar System, more than 500 'exoplanets' have been discovered orbiting other stars.

46. Why are planets round?

Gravity is a universal attractive force between all masses, so every chunk of a large body tries to pull every other chunk towards it.

If material can flow, body forms a sphere – the shape which ensures every constituent chunk is as close as possible to every other chunk.

Giant planets like Jupiter & Saturn are made of gas (and liquid, deep down where gas is compressed), which flows. They are therefore round.

Actually, Jupiter and Saturn have bulging waists. Because they are spinning fast, gas at their equators tends to be flung outwards.

Rocky & icy bodies are different. If small, gravity is too weak to squeeze interiors enough to flow. So, they are irregular like potatoes.

But the bigger a body, the stronger the force of gravity pulling its material together and squeezing it.

At a particular size, gravity is strong enough to make interior flow. For a rocky body, threshold ~400 km across; for icy body ~600 km.

Consequently, in the Solar System, all rocky bodies bigger than ~400 km across are round, as are all icy bodies bigger than ~600 km.

So, it's a battle between gravity, which crushes matter, & electromagnetic (EM) force, which makes matter stiff so it opposes gravity.

EM force by which electrons in neighbouring atoms repel each other is more than 1000 trillion trillion trillion x bigger force than gravity . . .

. . . So need large number of atoms clumped together – that is, a big astronomical body – before gravity wins.

Of course, if there's enough mass, gravity is overwhelming & nothing in universe can defy it. Result: black hole. But that's another story!

47. What is the smallest planet?

The smallest planet in our Solar System is Mercury. With a diameter of just 4880 km, it's only 40% larger than our Moon.

Mercury holds many planetary records: smallest, innermost, fastest, densest; biggest temperature range; most elongated & most skewed orbit.

Being so close to Sun (58m km), Mercury is visible just above the horizon at dusk or dawn. It's obscured if there are buildings or trees.

Inspection of surface markings through a telescope seemed to suggest Mercury keeps one face perpetually to the Sun.

But radar observations in the 1960s revealed Mercury rotates once every 59 days – 2/3 of the 88 days it takes to orbit the Sun.

Like the Moon, Mercury is cratered, its surface battered by impacts. Once, it had active volcanoes, but they died a few billion years ago.

Because of Mercury's high temperature and weak gravity (37% of Earth's), it has no atmosphere. Temperature: +450 °C (day) & -185 °C (night).

Remarkably, evidence for ice on Mercury. Planet's axis is not tipped, like Earth's. Means floors of deep craters at poles always in shadow.

Mercury's iron/nickel core is huge, compared to planet. If partially molten, with circulating currents, this may explain magnetic field.

Iron core may be big because, in distant past, Mercury was much bigger. Giant impact could have blasted away most of planet's rocky mantle.

First spacecraft to Mercury was *Mariner 10*: during three flybys in 1974 and 1975, mapped half of planet. Most craters named after artists.

In March 2011, NASA spacecraft *Messenger* arrived in orbit. Studies surface composition, magnetic field & internal structure of Mercury.

48. Why is Venus the closest place we know to hell?

Venus is brightest object in the sky after Sun and Moon. Named after Roman goddess of love because of its enchanting, beautiful appearance.

Because it's closer to Sun than Earth (108m km), Venus can be seen only after sunset (evening star) or before sunrise (morning star).

With diameter of 12,103 km, Venus is only slightly smaller than Earth. Internal structure – iron core, rocky mantle – probably very similar.

Venus appears not to have moving 'plates' like Earth. But surface is young, possibly due to global volcanism or other geologic upheaval.

Venus rotates backwards, as if planet is upside down. Also, its day (243 Earth days) is longer than its year (225 days). Not known why.

Super-high surface temperature of 500 °C revealed by *Mariner 2* – first interplanetary spacecraft – in 1962. Cause: strong greenhouse effect.

Thick atmosphere (mainly carbon dioxide): surface pressure 90 x Earth's. Surface hidden from view by thick clouds, containing sulphuric acid.

With temperatures high enough to melt lead, crushing surface pressure, acid clouds and violent lightning, Venus is Hell.

In 1970s, Soviet *Venera* landers parachuted down through clouds and photographed surface. Hostile environment quickly destroyed them.

1990–94, NASA's *Magellan* spacecraft orbited planet, mapping surface with cloud-penetrating radar. 'Saw' rolling plains, craters, volcanoes.

Since April 2006, European *Venus Express* has been in orbit, studying atmosphere/climate. Searching for active volcanoes with heat sensors.

All features on Venus are named after women (apart from Maxwell Mountains, named for 19th-C Scottish scientist whose work enabled radar).

49. Why is Mars red?

Mars, because of striking red colour, was named after Roman god of war. It orbits the Sun once every 1.88 years.

Mars, at 228m km from Sun, orbits outside Earth's orbit. When Earth overtakes it (every 26 months), Mars is visible all night long.

Through a telescope, Mars appears similar to Earth: dark surface markings, polar caps, tilted rotational axis, day of 24.6 hours.

Main difference: Mars is much smaller (6794 km). Surface gravity only 38% of Earth's. Can hang onto only very thin atmosphere, mostly CO_2.

While Mercury and Venus have no natural satellites and Earth has one, Mars has two: Phobos (27 km) and Deimos (15 km), discovered in 1877.

In 1972, *Mariner* 9 mapped Mars from orbit, discovering huge canyons, giant volcanoes, dry river beds, outflow channels and dune fields.

Valles Marineris (4000 km long, 6 km deep) is largest canyon in Solar System. Olympus Mons (500 km across, 25 km high) is tallest mountain.

Surface temperature on Mars varies between +10 and -80 °C. Most of planet is rocky desert, cold and arid, lashed by global dust storms.

Red colour of Mars is due to rust (iron oxide). The first colour pictures of the surface were taken by two *Viking* landers in 1976.

Many orbiters have studied Red Planet. Some still working: *Mars Reconnaissance Orbiter*, *Mars Odyssey* (both NASA) & *Mars Express* (ESA).

Also Mars rovers. *Spirit* and *Opportunity* landed in early 2004. *Opportunity* is still active. They found evidence of oceans and rivers in past.

Humans on Mars is a distant dream. More modest goal: return rocks from Mars to Earth, where they can be examined for fossil microorganisms.

50. Why are Venus, Earth and Mars so different?

Venus is closer to Sun than Earth; Mars is further away and smaller. But geologically speaking, the three planets are not too dissimilar.

Soon after formation, they probably all had clement temperatures, surface water, and rather thick atmospheres, with methane and ammonia.

Today, Venus, the Earth and Mars are very different. Venus boiled dry. Mars froze. Earth alone remained hospitable to life.

Volcanism on young Venus produced CO_2. Surface temperature rose because of heat-trapping by greenhouse effect; oceans began to evaporate.

Water vapour increased greenhouse effect. All liquid boiled away. Sun's UV light broke apart water molecules in atmosphere.

Resulting hydrogen escaped into space; oxygen bound to surface rocks. With no water cycle left, build-up of CO_2 in atmosphere continued.

If Venus had plate tectonics, like Earth, carbon-containing rocks would partly be recycled, so CO_2 build-up would have been slower.

However, lack of lubricating water may have stalled plate tectonics on Venus. This, and closeness to Sun, led to planet's hellish fate.

In contrast, Mars, being smaller than Earth, lost its heat faster. As its interior solidified, it died geologically, and its surface froze.

Without volcanoes to spew out the greenhouse gas CO_2, the planet lost its heat-trapping blanket. The temperature plummeted. Water froze.

With no magnetic field to protect it, and only weak gravity to hold onto it, Mars's thinning atmosphere was stripped away by the solar wind.

If Mars had been larger (more heat, stronger gravity), or had orbited closer to the Sun, it might have escaped its frigid fate.

If closer to Sun, Earth would have boiled dry, like Venus. If much smaller, Earth would have frozen and lost its atmosphere, like Mars.

Venus shows that too much CO_2 makes a planet very hot. Mars shows that too little CO_2 makes it very cold. Both stand as warnings to us.

On Earth, everything is 'just right' for the existence of water and life. We are fortunate indeed to live on the 'Goldilocks planet'.

51. Is there water on Mars?

Lots of it. But all frozen. Most water is locked up in subsurface ice at high latitudes. The polar caps also contain large amounts of ice.

Late 19th C, Giovanni Schiaparelli observed straight lines on Mars. Called them *canali* (channels). Lost in translation, became 'canals'.

Percival Lowell thought they were artificial waterways, built by Martians to irrigate dry equatorial regions with water from polar caps.

Martian canals were visual illusion. However, space probes later found dry runoff and outflow channels, indicating ancient water on Mars.

NASA's *Viking* orbiters confirmed existence of water on Mars. Polar caps are mainly frozen CO_2, but they contain significant water ice too.

Subsurface ice, probably permafrost, was revealed by *Mars Odyssey* (neutron spectrometer data) and *Mars Express* (radar measurements).

Subsurface ice was actually scooped up by *Phoenix* lander, which touched down in arctic region. Remaining Q: did Mars ever have liquid water?

Channels and other river-related features suggest it did – a few billion years ago, when the atmosphere was thicker and the planet warmer.

In fact, minerals formed in water and found by the rover *Opportunity* confirm that, in the remote past, there were lakes and seas on Mars.

Mars was once a waterworld, with an extensive ocean covering most of the low-lying northern hemisphere. It may have looked just like Earth.

Water-cut gullies on the inner slopes of craters suggest that, even today, subsurface ice may occasionally melt and flow across surface.

Unfortunately, with Mars's atmospheric pressure being only 0.7% of Earth's, any water unleashed onto the surface will evaporate immediately.

52. Was the Mars Face built by an alien civilisation?

You might as well ask if Ayer's Rock (Australia) was built by aliens. The 'Mars Face' looks peculiar, but it's just a natural formation.

Discovered in a grainy *Viking* orbiter photograph, it's an isolated hill that vaguely resembles a human face with mouth, eyes & nostrils.

The illusion is created by light and shadows – and picture defects. Later, sharper photos revealed the Mars Face to be a mesa formation.

Still, some believe NASA is covering up evidence for existence of lost Martian civilisation. And it's impossible to convince them otherwise.

Intelligence on Mars – often green & slimy – is a major theme of science fiction. Take, for instance, H. G. Wells's *The War of the Worlds*.

Until 1960s, astronomers imagined lichens or simple plants on Mars. Until recently, Arthur C. Clarke even believed in Martian banyan trees.

Biology experiment on NASA's 1976 *Viking* landers found strange soil chemistry. Experiment designer still claims it found Martian microbes.

Mars's warm, wet history means life might have arisen in past. Future missions to Red Planet will look for evidence of Martian microbes.

Some microorganisms may even have survived until today in subsurface pockets of water, shielded from deadly cosmic rays and solar UV.

A tell-tale sign of life is methane. And that gas, surprisingly, has been spotted on Mars. May point to presence of life there today.

If life did arise on Mars in distant past, microbes could have travelled to Earth inside Martian meteorites, flung into space by cosmic impacts.

If so, life on Earth may have originated on Mars. In a sense, your best chance to see a real Mars Face may be to look in a mirror.

53. How dangerous is a flight through the asteroid belt?

Over half a million asteroids are known. Most orbit the Sun between the orbits of Mars and Jupiter. Sounds like a crowded/dangerous place.

But don't believe SF movies. Average distance between asteroids is comparable to Earth–Moon distance. Asteroid belt is mainly empty space.

Also, most asteroids are small. Only 200 are larger than 100 km across. The total mass of all asteroids is a mere 4% of the mass of the Moon.

Nevertheless, collisions in asteroid belt do occur, creating families of smaller objects with similar orbits and identical composition.

Since 1973, several space probes have flown through the asteroid belt without problems. Some have even been on missions to asteroids.

First asteroid, Ceres, discovered by Giuseppe Piazzi in 1801. Was hailed as 'missing planet' predicted for gap between Mars and Jupiter.

But within a few years, 3 more 'planets' found in same region: Pallas, Juno, Vesta. By 1840s, number of planets in Solar System reached 11.

Soon, hundreds of bodies turned up. Astronomers realised Ceres is just largest member of new class of objects: asteroids, or minor planets.

Ceres, at 975 km across, is only asteroid large enough to have become spherical through its self-gravity. Officially a 'dwarf planet'.

Most asteroids are lumpy rocks or porous conglomerates of pebbles and dirt. Some are binary. Many have one or even two small moons.

Asteroids visited by spacecraft: Gaspra, Ida (with moon, Dactyl), Mathilde, Braille, Eros, Annefrank, Itokawa, Steins, Vesta and Lutetia.

NASA's *Dawn* spacecraft, launched in September 2007, entered orbit around Vesta in July 2011. Will be moving on to Ceres, arriving 2015.

Asteroids are leftovers from formation of Solar System. Comparable to planetesimals and protoplanets from which terrestrial planets formed.

54. Did a killer asteroid wipe out the dinosaurs?

Most asteroids circle Sun in the asteroid belt, between the orbits of Mars and Jupiter. Such 'main belt' asteroids pose no risk to Earth.

However, due to collisions and/or gravitational nudges from Jupiter, some asteroids enter inner Solar System, and may cross Earth's orbit.

These near-Earth objects (NEOs), which also include comets, may hit Earth, causing global devastation. Has happened; will happen again.

Tunguska explosion in Siberia in 1908 flattened 2000 square kilometres of forest. Caused by comet fragment size of couple of houses (30 m).

Arizona's Meteor Crater, more than a kilometre across, was created by the impact of iron meteorite the size of a terrace of houses (50 m).

Larger impacts are much less frequent. 1-km asteroids strike Earth every 500,000 years or so. 10-km or larger, once every 100m years.

Dinosaurs and many other species became extinct 65m years ago. Around that time, Earth was hit by comet or asteroid 10 m across.

Extinctions not necessarily caused by impact. Could be due to extreme outpouring of volcanic lava that created 'Deccan Traps' in India.

Still, NEOs pose real danger. Would create global fires/ tsunamis. Dust in stratosphere would block sunlight to surface for years.

Dedicated survey telescopes scan skies for potentially hazardous objects (PHOs). Goal is to detect most PHOs larger than 140 m before 2020.

If found, killer asteroid's orbit could be changed by rocket power, nearby nuclear explosion, or gravitational tug from massive spacecraft.

Dinosaur extinctions had silver lining. Left abundant niches for mammals to fill. Without the killer asteroid, we probably wouldn't be here.

55. Is Jupiter a failed sun?

In the movie *2010: Odyssey 2*, aliens turn Jupiter into a second sun to give helping hand to nascent life on Jupiter's ice moon, Europa.

But is Jupiter really a failed sun? How close did it come to igniting sunlight-generating nuclear reactions & blazing in Earth's skies?

Key fact 1: to trigger fusion of hydrogen into helium – power source of Sun – requires central temperature of more than 10 million degrees.

Key fact 2: The gravity of a giant ball of gas squeezes it, heating it. The bigger the mass, the bigger the squeezing and the hotter it gets.

To attain a temperature of about 10 million degrees requires a mass equivalent to about 8% of the mass of the Sun, or 80 Jupiters.

So Jupiter failed by quite a large margin to be second sun. But there is a twist . . .

Jupiter gives out more than twice as much heat as it receives from Sun (core is slowly shrinking, converting gravitational energy to heat).

So Jupiter does not conform to strict definition of planet: body that shines not from its own light/heat but from that reflected from a star.

So how did aliens in *2010* make Jupiter a sun? Since gravity sucks from within, & not up to job, they must have compressed from without.

Jupiter is the 5th planet from the Sun. At 143,000 km across at the equator, it is the largest planet in the Solar System.

Jupiter orbits the Sun every 11.86 years. Despite being 778.5m km from the Sun, it's brightest planet in night sky apart from Venus.

56. Does Jupiter change appearance?

Jupiter is a gas planet. Basically, it's all atmosphere. What you see from outside is cloud structures. Very dynamic and ever-changing.

Clouds are stretched into belts and bands by fast rotation of planet: once every 9h 55m, corresponding to 45,300 km/h at equator.

Small telescope reveals two dark cloud belts on either side of equator. But even those aren't permanent: south belt disappeared in 2010.

Great Red Spot (GRS) is also prominent in small telescope. It's a giant anticyclonic storm system, partly embedded in south equatorial belt.

GRS was first described by Robert Hooke in 1664. Has been observed continuously since first half of 19th century. Size and colour vary.

Average dimensions: 30,000 x 13,000 km, or twice as large as Earth. Colour: usually salmon, but changes from very pale to strong orange/red.

GRS rotates anticlockwise in 6 days. Cloud tops in centre are cooler and ~8 km higher than surroundings. Wind speeds up to 450 km/hour.

Origin of red colour unclear. Maybe due to organic phosphorus or sulphur compounds dragged up from lower levels in Jupiter's atmosphere.

Larger telescopes/space probes also reveal smaller spots: white ovals. Some merge and grow over time. In the future, new Red Spots may form.

In July 1994, fragments of comet Shoemaker-Levy 9 crashed into Jupiter's atmosphere, creating temporary dark blemishes that faded over time.

In recent years, amateur astronomers have detected additional temporary dark spots, and sudden bright flashes, probably related to impacts.

To sum up: A Jupiter atlas makes no sense, as the planet never looks exactly the same. But most changes are hard to see in small telescope.

NASA's *Juno* spacecraft, due to arrive at Jupiter in July 2016, will study ever-changing atmosphere in detail.

57. What is special about Jupiter's moons?

The 4 giant Jovian moons are visible in a small telescope & were discovered by Galileo from Padua in 1610. Christened 'Galilean moons'.

Discovery revealed another body around which other bodies circle. Fatally undermined Church-supported idea that Earth is centre of cosmos.

Jupiter's giant moons allowed first accurate estimate of speed of light, even though a million times faster than speed of sound.

In 1676, Ole Christensen Rømer noticed moons went behind Jupiter 22 minutes later when Earth on the far side of Sun to Jupiter.

22 mins = time light takes to cross Earth's orbit. Knowing distance, Rømer estimated light speed at 225,000 km/s. Modern value 300,000 km/s.

One Galilean moon, Io, is hottest body in the Solar System. Remarkably, it generates more heat per unit volume than even the Sun.

If you continually squeeze a squash ball it gets hot. Io is hot for same reason. Only Jupiter's enormous gravity is doing the squeezing.

Io is most active body in Solar System. Interior molten. Surface peppered with volcanoes. Pumping 10 bn tonnes of matter a year into space.

Actually, Io has geysers not volcanoes. What is erupting, hundreds of kms into space, is not lava but super-heated sulphur dioxide gas.

Second giant moon, Europa, is biggest ice rink in Solar System. Biggest ocean in Solar System may not be on Earth but beneath Europa's ice.

Europa, like Io, is squeezed & stretched by Jupiter's gravity. Heat melts its icy interior. Under 10 km layer of ice, ocean 100 km deep.

Europa is most exciting body in Solar System. Though ocean is totally dark, as on Earth life could exist around volcanic vents on sea floor.

Third Galilean moon, Ganymede, is Solar System's largest satellite. At 5262 km across, it is even bigger than the planet Mercury.

Outermost large moon, Callisto, is only one beyond Jupiter's deadly radiation belts. If humans explore Jovian system, it will be the perfect base.

58. Could Saturn float in water?

Like Jupiter, Saturn is a giant gas planet. But it's smaller: 120,500 km. At 1.4 bn km distance, Saturn takes 29.5 years to orbit the Sun.

In ancient times, Saturn was most distant planet known, with slowest motion. Named after Roman god, son of Earth and sky, father of Jupiter.

Saturn is best known for its spectacular ring system, visible in small telescope. First seen (but not recognised as such) by Galileo, 1610.

True nature of rings was discovered in 1655 by Dutch scientist Christiaan Huygens, who also discovered Saturn's largest moon, Titan.

Saturn, unlike Jupiter, has tilted axis (27°) and seasons. Ring shadow on winter hemisphere adds to seasonal temperature variations.

Saturn, being further from Sun, is colder than Jupiter. So clouds form deeper in atmosphere; are less conspicuous, making planet look bland.

Exceptions: 1933 Great White Spot, seen by English comedian/amateur astronomer Will Hay. More recently: giant storm, December 2010.

Saturn, at its north pole, has mysterious 25,000-km-wide hexagonal hurricane. Discovered by *Cassini* spacecraft, in orbit since 2004.

Saturn's fast rotation – once every 10h 39m – makes waist bulge outwards. Polar diameter is 90% of equatorial. Winds as fast as 1800 km/h.

Saturn consists mainly of light gases: hydrogen/helium. Planet is 95 times mass of Earth. Average density is a mere 0.69 gram/cm^3.

Density comparable to that of elm wood. So, if you could find a large enough body of water, Saturn would float in it.

59. How thin are Saturn's rings?

The answer is: incredibly thin. Saturn's rings, despite stretching more than 100,000 km from inner to outer edge, may be only 20 m thick.

Put it another way: if the rings were shrunk to 1 km in diameter, they would be thinner than the sharpest razor blade.

Galileo was giant in history of science. Realised swing of pendulum is regular & all bodies fall same rate. But low point in career was . . .

. . . when in 1610 he turned his newfangled telescope on Saturn and declared it was . . . 'a planet with ears'.

The next year Galileo decided Saturn had two big moons – one on either side. But the moons later vanished. Galileo died baffled.

Mystery solved only in 1655 when Christiaan Huygens viewed Saturn through bigger telescope & deduced planet is girdled by system of rings.

As Saturn orbits Sun, rings change orientation as seen from Earth. When edge-on, they vanish. When at angle, they do indeed look like ears.

In 1858, James Clerk Maxwell proved that, if rings solid or fluid, could not be stable. Must be shoal of independently orbiting particles.

Particles are 99% water ice, which explains brightness of rings. Although typical size 1 cm, particles range from dust grain up to house.

Rings believed to be about 400 million years old and to have formed when a 250-km icy moon wandered too close to Saturn and was shattered.

In early 1980s, NASA's *Voyager* spacecraft discovered that Saturn's rings in fact consist of thousands upon thousands of narrow ringlets.

Actually, Saturn does not have rings at all – it has multiple spirals, like the grooves on an old vinyl record.

Vibrations of rubble cause 'spiral density wave' to move outwards. As passes, compresses together particles, making 'temporary' ringlets.

Spiral density wave also creates 'spiral arms' of Milky Way. Amazingly, Saturn's rings are just a tightly wound version of a spiral galaxy.

60. Could I swim on Titan?

Titan, discovered by Christiaan Huygens in 1655, is largest moon of Saturn and second largest in Solar System. Larger than planet Mercury.

In 1944, Gerard Kuiper detected methane on Titan. Definitive proof of atmosphere. Titan is only planetary moon with substantial atmosphere.

Voyager spacecraft found: atmosphere is very thick. Main constituent is nitrogen. Haze layers of organic molecules hide surface from view.

Titan's atmosphere resembles that of primitive Earth. Surface pressure: 1.45 x Earth's. Temperature, however, much lower: -180 °C.

On 14 Jan 2005, European *Huygens* probe soft-landed on Titan. Found that ice plays role of rock; liquid methane plays role of water.

Cassini spacecraft mapped large parts of Titan using radar. Found lakes of liquid methane and ethane, and evidence for methane rain storms.

Except for Earth, Titan is only world in Solar System with surface liquid. So yes, you could swim there. Would be very unhealthy, though.

Small and icy Saturnian moon Enceladus (500 km) is interesting too. Has geysers of water, ice and dust, indicating subsurface ocean.

Small inner moons of Saturn sculpt ring system with their gravity. Best example: 4700-km-wide Cassini gap carved by gravity of Mimas.

Many smaller ring structures, such as narrow F-ring and smaller gaps, caused by gravity of small neighbouring or embedded moonlets.

Iapetus is strange Janus-faced moon, with dark and bright hemispheres. Giant equatorial ridge possibly caused by impacting ring material.

Little Phoebe (230 km) is probably a 'centaur': Kuiper Belt Object captured by Saturn's gravity from beyond orbit of Neptune.

Hyperion (328 km) is weird, extremely porous object, mainly consisting of frozen water. Resembles sponge: about 40% of volume is empty space.

61. Why is Uranus lying on its side?

Since all planets were born from disc swirling around newborn Sun, all should be spinning upright, with equators in plane of their orbits.

Two exceptions: Venus, which is upside down, rotating in opposite sense to orbital direction; Uranus, which is spinning on its side.

As Uranus orbits Sun every 84.3 yrs, north pole points towards Sun & gets 42 years of sunlight, then points away & has 42 years of darkness.

Q: Why is Uranus like top that's fallen over? A: Perhaps knocked over by collision with big body (Earth's Moon created in similar smash-up).

Problem: Uranus's moons are tilted with the planet, all orbiting its equator. Hard to imagine impact tilting both Uranus and its moons.

In 2009, Gwenaël Boué and Jacques Laskar at the Paris Observatory proposed an alternative theory.

Gravity of debris disc swirling around newborn Sun could have made spin of embryonic Uranus wobble, or 'precess', like a spinning top.

If the planet once had a giant moon, with 0.1% the planet's mass, wobble might eventually have got so wild it tipped planet on its side.

But where is the giant moon? Boué & Laskar say it was stolen! Specifically, they say friction between the protoplanetary disc & Uranus . . .

. . . caused the planet to 'migrate' through the disc. As it did so, it passed another giant planet, whose gravity grabbed the moon.

Might seem far-fetched. However, it has long puzzled astronomers that Uranus is only 1 of Solar System's 4 giant planets without a big moon.

Incidentally, Uranus was first planet found unknown to ancients. William Herschel discovered it from his garden, Bath, England, 1781.

Herschel, a German immigrant, named planet 'George's star' after King George III. French objected. Germans suggested Uranus.

Herschel's discovery doubled size of the Solar System. Uranus, which has 4 x diameter of Earth, orbits about 20 x as far from Sun as Earth.

Uranus, despite being on its side, is pretty dull and featureless. It gets most astronomers' vote as 'most boring planet'.

62. Has Neptune always been the outermost planet?

Neptune was actually seen and recorded by Galileo Galilei in 1612. He spotted it close to Jupiter but mistakenly thought it was a star.

Existence of planet inferred from wayward orbit of Uranus. Being tugged by unseen planet. Location calculated from Newton's law of gravity.

Calculations carried out in mid-1840s by John Couch Adams in England (imprecise) and Urbain Le Verrier in France (much more accurate).

On 23 Sep 1846, Johann Galle in Berlin found missing planet close to position predicted by Le Verrier. Named Neptune after Roman god of sea.

First (and only) spacecraft to visit Neptune was NASA's *Voyager 2*, on 25 Aug 1989. Discovered new moons & dark, narrow rings around planet.

Neptune, like Earth, is a blue planet. Colour due to atmospheric methane. Planet has strong winds, giant storms & cirrus-like wispy clouds.

Neptune, at 49,530 km across, is smallest of 4 giant planets. Spins once every 16h 07m. At 4.48 bn km from Sun, takes ~165 years to orbit.

Largest moon Triton found by Lassell, 1846. Orbits opposite way to Neptune's spin. Probably captured Kuiper Belt Object, similar to Pluto.

Neptune is currently the outermost planet in the Solar System. However, it has not always been so – for three reasons . . .

1) In antiquity, Saturn was most distant planet known.
2) Until August 2006, Pluto, further from Sun, considered a planet (Pluto found 1930).

3) Computer simulations show that pull of 'migrating' Jupiter may have caused Uranus and Neptune to swap places after birth of Solar System.

63. Why is Pluto no longer considered a planet?

Taking into account Neptune's gravity, motion of Uranus still seemed erratic. US astronomer Percival Lowell began search for 9th planet.

After Lowell's death, Lowell Observatory hired Kansas farm boy Clyde Tombaugh to expose and search photographic plates for 'Planet X'.

On 18 Feb 1930, Tombaugh triumphant. Found new object on plates taken several weeks earlier. Discovery of Planet X announced 13 Mar 1930.

'Pluto' proposed by 11-year-old Venetia Burney of Oxford, England. Roman god of underworld. First two letters initials of Percival Lowell.

Pluto's orbit is weird. Very inclined (17°) to orbital plane of other planets, and very elongated: varies between 4.4 & 7.4 bn km from Sun.

Pluto fainter/smaller/less massive than expected for 9th planet. Mass, deduced from moon, Charon, discovered in 1974, mere 18% of Moon's.

Since 1992, hundreds of icy objects found beyond Neptune's orbit. Dawned on people Pluto is merely very large member of this Kuiper Belt.

In fact, Eris, another Kuiper Belt Object (KBO), may be slightly larger than Pluto. The 2300-km-diameter body even has a moon, Dysnomia.

Some KBOs have extremely wide/inclined/elongated orbits. Sedna, for instance, takes 12,000 yrs to orbit Sun, compared with Pluto's 248 yrs.

KBOs are icy leftovers from Solar System's formation. Total number larger than 100 km is now estimated to be about 100,000.

Pluto's status changed at Prague meeting of International Astronomical Union, August 2006. Downgraded from planet to 'dwarf planet'.

Fortunately, Clyde Tombaugh did not see the ignominious demotion of his beloved planet. He died, aged 90, in 1997.

NASA's *New Horizons* probe, launched in Jan 2006, will fly by Pluto and Charon in June 2015, before heading for at least two other KBOs.

64. What are comets?

Comets were believed to be stars-with-tails, or hairy stars. Name comes from Latin for hair – *coma*. Can be visible for weeks in night sky.

Aristotle thought comets were vapours glowing in atmosphere. Tycho Brahe realised they were further away than the Moon and cosmic in nature.

Edmond Halley realised 1682 comet followed same orbit as comets of 1531 & 1607. All same object! Predicted return of Halley's Comet in 1758.

Comets travel around Sun in highly elliptical orbits. Periods vary wildly, from a few yrs to thousands of yrs. Halley's Comet takes 76 yrs.

Comets are porous chunks of ice/dirt few kms across. Primitive material left over from birth of Solar System, so scientifically priceless.

When close to Sun, ice evaporates, dust particles lost. Comet grows bluish gas tail and yellowish dust tail, pushed like windsocks from Sun.

Tail, though extremely tenuous, can be spectacular. Dust spreads around orbit. Enters atmosphere as meteors if Earth crosses comet's path.

When Solar System formed, trillions of comets born on periphery. Many incorporated into icy cores of giant planets, or Kuiper Belt Objects.

However, close encounters with embryonic giant planets flung most comets into distant 'Oort Cloud', source of today's long-period comets.

Small comets often crash into Sun or Jupiter, as did Comet Shoemaker-Levy 9 in 1994. Others eroded away by solar heat after many orbits.

In early days of Solar System, impacting comets may have ferried most of water on Earth. Flipside: impacts have caused mass extinctions.

A few comets have been studied close-up and even sampled by spacecraft. European *Rosetta* craft will land *Philae* probe on a comet in 2014.

65. Where is the Solar System's edge?

The Solar System does not have a sharply defined edge. It's like asking: where's the edge of the Rocky Mountains?

If Solar System defined as just Sun & planets, edge is 4.5 bn km from Sun (distance of Neptune). However, Solar System contains much more.

Small icy bodies of Kuiper Belt extend from just beyond Neptune to 7 bn km from Sun. But despite belt's abrupt edge, some travel further.

For instance, 1500-km Sedna, discovered in 2003, travels out to 143.7 bn km from Sun in its highly elongated orbit.

Oort Cloud of cometary 'nuclei' even extends out to about 1 light year (9.46 trillion km). That's 25% of way to nearest star.

In 1950, Dutch astronomer Jan Oort deduced from orbits of long-period comets that they came from vast reservoir far from Sun: Oort Cloud.

Oort Cloud may contain few trillion comets larger than 1 km. Despite huge number, however, their average separation is at least 1 bn km.

Oort Cloud more or less defines outer edge of sphere of influence of Sun's gravity. Widely considered to be part of Solar System.

If a comet is nudged by gravity of another comet/passing star, it may be knocked into orbit taking it towards Sun as a long-period comet.

Gravity of giant planet like Jupiter can trap long-period comet in the inner Solar System, turning it into short-period comet like Halley's.

Another type of Solar System edge, ~15 bn km from Sun, is defined as the limit of the 'heliosphere' – Sun's magnetic sphere of influence.

Most particles in heliosphere come from solar wind, which carries Sun's magnetic field. Beyond the heliospere is interstellar space.

Heliosphere is tear-shaped because of Sun's motion through Galaxy. *Voyager* spacecraft are expected to leave heliosphere around 2014.

The Stars

66. What are stars?

Stars are other suns, shrunk to mere pinpricks of light by their mind-bogglingly huge distance from the Earth.

In 1600, for maintaining that the stars are other suns, the Italian philosopher Giordano Bruno was burnt at the stake by the Catholic Church.

A star is a giant ball of gas – almost entirely made of hydrogen & helium, the 2 lightest elements – held together by its own self-gravity.

The core of a star is squeezed so much by the weight of material bearing down from above that it is heated to more than 10 million °C.

Super-high temperatures trigger nuclear reactions which initially 'fuse' hydrogen into helium. Their byproduct is heat/sunlight.

The distinction between a star and a planet is that a star generates its own heat & light while a planet shines merely by reflected light.

A star's luminosity (& how quickly it squanders its nuclear fuel) is determined by its mass. Massive stars burn brightly & have short lives.

Our galaxy contains more than 100,000,000,000 stars. In the universe there are 10,000,000,000,000,000,000 (give or take a few).

About 6000 stars are visible to the naked eye. Almost all are hugely more luminous than the Sun, which itself is brighter than the average.

Paradoxically, most nearby stars are invisible to the naked eye. They are cool, faint 'red dwarfs', which account for ~70% of all stars.

Red dwarfs are so miserly with the burning of their nuclear fuel that many will live for 10 trillion years, 1000 times as long as the Sun.

The nearest star is of course the Sun. Its light takes 8.3 minutes to get here. Next nearest star is Alpha Centauri, 4.2 light years away.

Alpha Centauri is in fact a triple star system. Most stars are actually doubles or triples. The Sun is an exception – a rare lone star.

One of the central goals of astronomy is to look back far enough in time to see the universe's first stars as they started to switch on.

67. Why do stars twinkle?

'Twinkle, twinkle, little star, how I wonder what you are. Up above the world so high, like a diamond in the sky,' wrote Jane Taylor in 1806.

Ancient people noticed that stars twinkle whereas planets do not. They also noticed stars appear fixed to firmament whereas planets wander.

Both phenomena are explained by distance. Stars are so immensely far away that they appear pinpricks and their movement is unnoticeable . . .

. . . Planets, on the other hand, are relatively nearby so they appear as tiny discs in a telescope, and their movement across sky is marked.

Looking up through turbulent atmosphere at stars & planets is like looking at lights on a swimming pool ceiling from bottom of pool.

Wavering of water makes point-like lights jitter back/forth (twinkle). But bigger lights merely ripples around edges, so stays steady.

Similarly, stars twinkle because they are small compared with 'clumpiness' of atmosphere; planets remain steady because they are big.

Twinkling of stars smears out images in telescopes. One way to get sharper images is to go above atmosphere (Hubble Space Telescope).

Another way to compensate for twinkling is to flex the shape of a thin telescope mirror many times a second (adaptive optics).

Pinprick sources of radio waves such as 'pulsars' also twinkle (interstellar scintillation) due to turbulence of gas between the stars.

68. How do we know the distance to the stars?

If an object appears to move a lot when seen from two separated vantage points, it must be close; if it moves only a little, it is far away.

See for yourself. Hold your finger close & observe it with 1 eye, then the other. It moves a lot. Do same with it far away. Moves a little.

This effect (parallax) can reveal distance of star. Observe stars from two points on opposite sides of Earth's orbit (six months apart).

Star said to be at distance of 1 parsec (3.26 light years) if, when observed six months apart, direction changes by 1 arcsecond (1/3600°).

Problem: turbulence in atmosphere smears star images to 0.5 arcsecond or more, so parallax can reveal distances of only nearest stars.

Solution: go into space. European *Hipparcos* satellite used parallax to pin down distances of 100,000 stars out to more than 100 light years.

To measure larger distances, you must identify stars with known intrinsic luminosity. If one star fainter than another, it is more distant.

Are there stars whose intrinsic luminosity is known? Yes. 'Cepheid variables' – super-luminous stars that pulsate like beating hearts.

Crucial discovery of Henrietta Leavitt in 1912: Cepheids that are intrinsically more luminous vary their brightness over longer time period.

To find distance to Cepheid: 1) time 'period' light variation → intrinsic luminosity; 2) compare intrinsic & apparent brightness → distance.

In 1923, Edwin Hubble observed Cepheids in Andromeda nebula, finding it was an island 'galaxy' far from Milky Way (2.5 million light years).

NASA's Hubble Space Telescope has spotted Cepheids in galaxy M100, pinning down stellar distances out to 56 million light years from Sun.

69. How do we know what the stars are made of?

In 1835, philosopher Auguste Comte claimed self-evident that one thing science will never know is composition of the stars. He was wrong.

Nature is kind to us. Atoms of each element emit light of characteristic colours/wavelengths, enabling identification of elements in stars.

A unique 'spectral' fingerprint exists because each atom of a particular element has unique arrangement of orbiting electrons.

When an electron jumps from one orbit to another, light is given out. Energy of light equal to the difference in energy between two orbits.

Complication: Stars are so hot some atoms have had most or all electrons smashed away. So an element, even though common, may not show up.

Before this was realised, people were hoodwinked into thinking Sun was made of iron since iron has the most prominent spectral fingerprint.

Breakthrough made by Cecilia Payne in 1925. From sunlight, she deduced that hydrogen & helium – 2 gases rare on Earth – make up 98% of Sun.

Payne had stumbled on composition of the universe. 98% of all atoms in the cosmos are hydrogen & helium. All the rest account for a mere 2%.

Despite writing most important astronomy PhD of 20th C, Payne is hardly known. Suffered from being woman scientist in male field.

People gradually realised elements are present in same proportions everywhere. This implies elements are made by universal process.

But where is the 'furnace' that forged the elements in our bodies? Finger was pointed at stars, then Big Bang, then back to stars again.

Fred Hoyle and colleagues, in a monumental 1957 paper, spelled out precise 'nuclear' processes that have built up the elements inside stars.

Problem: stars cannot make amount of helium seen in universe. Nature not simple. Heavy elements are made stars; light elements in Big Bang.

Hoyle's colleague, Willy Fowler, received 1983 Nobel Prize for physics for pinpointing origin of elements. Scandalously, Hoyle was ignored.

70. Are all stars single like the Sun?

Sun is actually unusual in being a lone star. More than half of stars in Milky Way are in multiple systems – two, three or even four stars.

In fact, closest star system to the Sun – Alpha Centauri system, 4.2 light years away – consists of 3 stars (Proxima Centauri closest).

In optical binary, we see two stars orbiting each other. In spectroscopic binary, light reveals spectral fingerprint of two stars.

Nobody knows why most stars are multiple. Must be telling us about processes in interstellar clouds (nurseries) where stars are born.

Was once thought hard for planets to exist in multiple systems. We now know, if two stars close, planets can exist in 'circumbinary' orbit.

If life exists elsewhere in the Milky Way, it may be most ETs live on planets with two or more suns burning down from the sky.

In 1984, David Raup & John Sepkosky actually suggested the Sun might have a super-faint companion – in a super-long 27-million-year-orbit.

Companion star, dubbed 'Nemesis', was proposed to explain 27m-year 'periodicity' found in mass extinctions in palaeontological record.

Every 27m years, claimed scientists, Nemesis shakes up comet cloud surrounding Solar System, sending comets Earthbound, causing die-offs.

Nemesis was never found. Anyhow, wouldn't have worked since nudges from gravity of nearby stars would make period of orbit fluctuate.

It is always possible, however, that long ago in the stellar nursery where it was born, the Sun had a brother, stolen away by passing star.

71. How do stars work?

A star is a giant ball of gas. It forms when an interstellar cloud of mostly hydrogen and helium begins shrinking under its own gravity.

The shrinkage continues until the core becomes so squeezed and hot it triggers 'nuclear fusion' of hydrogen into helium. Byproduct: heat.

Hot gas, pushing outward, stops gravity in its tracks. The ball shrinks no further. No longer a gas ball but a glowing gas ball: a star.

Key fact: nuclear fusion is hugely sensitive to temperature. If temperature goes up, it is boosted; if goes down, it is throttled back.

So, if heat generation drops, core shrinks/heats, boosting fusion; if heat generation climbs, core expands/cools, throttling back fusion.

Consequently, a star has a natural in-built thermostat. Keeps it perpetually balanced between shrinking and expanding.

Nothing is forever. Fusion reactions change hydrogen into helium, which, being heavier, sinks to centre, squeezing & heating up the core.

The internal structure of a star therefore changes gradually. The star evolves. And, sooner or later, its stable balance is upset.

A low-mass star like Sun evolves into a profligate red giant as it runs out of hydrogen. Then it dies a slow death as a fading white dwarf.

A high-mass star evolves ever more extreme conditions, triggering new fusion reactions, which achieve a series of new stable balances.

But gravity never goes away. Each new balance is short-lived. A star may win a few battles against gravity. But it can never win the war.

Eventually, gravity crushes the core to a black hole or to a ball of neutrons. This triggers a catastrophic explosion – a 'supernova'.

72. Are we made of stardust?

The universe, with its black holes, nebulae & exploding stars, seems unconnected with our lives. Nothing could be further from the truth.

The iron in your blood, calcium in your bones, oxygen that fills your lungs … all forged inside stars that died before the Earth was born.

The stars are furnaces in which heavy elements like copper are gradually built-up from nature's simplest Lego building block, hydrogen.

As stars turn one element into another, this alters both their chemical and overall structure. Their interiors change and they 'evolve'.

So the elements contain the key to unlocking secret of the stars and the stars contain the key to unlocking the secret of the elements.

Element-building proceeds furthest – all the way to iron – in most massive stars. Instability then causes them to explode as supernovae.

Elements even heavier than iron – such as uranium – are forged in the orgy of nuclear reactions in inferno of a supernova explosion itself.

In the cataclysmic violence of a supernova explosion, the products of the stellar furnace are scattered to the winds of space.

Supernova debris mingles with gas in interstellar clouds. When stars congeal from clouds, they are therefore enriched by heavier elements.

Each successive generation of stars has more heavy elements. Sun is reckoned to be 3rd-generation star – 2 generations died before was born.

But while heavy elements are built up inside stars, lightest element like helium were created in Big Bang fireball (first 10 minutes).

In fact, abundance of helium in universe – 10% of all atoms – is exactly what theory predicts. Powerful evidence universe began in Big Bang.

Astrologers are guilty not of being crazy but of not being crazy enough. We are far more connected to the stars than they ever imagined.

Want to see a bit of a star? Hold up your hand. You are stardust made flesh. You were literally made in heaven.

73. What are the differences between stars?

Given that the recipe for a star is so simple – ball of gas held together by its own gravity – it is remarkable that stars are so diverse.

Some stars live 10 trillion years – 1000 times longer than current age of universe – whereas others blow up after a few million years.

A star's mass determines its life expectancy. Massive stars are hot, so burn their fuel at breakneck speed. Low-mass stars barely smoulder.

Some stars are no bigger than Mount Everest (Crab pulsar) whereas others are big enough to swallow 10 billion Suns (VY Canis Majoris).

If VY Canis Majoris replaced the Sun it would swallow all the planets out to the orbit of Saturn, the 6th most distant planet from Sun.

Some stars are ferociously blue-white in colour, some are yellow-white (like the Sun) and some are dull cherry red, fading to black.

A star's temperature determines its colour. Blue-white stars super-hot (some more than 100,000 °C); red stars cool (a few thousand °C).

Some stars are constant whereas others pulsate in brightness/ even explode. Instability in massive stars caused by dodgy nuclear reactions.

Some stars rich in heavy elements like iron whereas others not. These can affect their structure/appearance by damming up heat inside.

How long ago star was born determines composition. Oldest stars formed before supernovae enriched galaxy with products of nuclear fusion.

Some stars have planets whereas some do not. (Since at least 10% do and each has several planets, there could be as many planets as stars.)

What determines whether stars have planets is not yet clear. However, it seems that heavy elements may be needed to make rocky planets.

Given that there are about 10,000,000,000,000,000,000 stars in the universe, what other oddities are there in the stellar zoo?

74. Why do stars explode?

Most stars, like the Sun, burn hydrogen to helium. But they never get dense/hot enough to go to the next step – burning helium to carbon.

So, most stars run out of H-fuel, have brief last gasp as swollen, profligate red giant, then slowly fade into invisibility as white dwarf.

Massive stars are different. After turning one element into a heavier one, they always get dense/hot enough to go to the next step.

Most massive stars end up 'silicon burning' – super-fast orgy of nuclear element building that eventually converts the core to iron/nickel.

An iron/nickel core spells disaster. Further element-building needs energy rather than creates it, sucking heat vampire-like from star.

Unable to generate heat to support its gas against the overwhelmingly powerful gravity trying to crush it, the core 'implodes'.

Implosion stopped only with formation of 'neutron core' – super-dense ball of neutrons. So hard that in-falling star bounces off it.

Implosion reversed into explosion (supernova). 'Neutrinos', subatomic particles created in birth of neutron core, blow off envelope of star.

A supernova can briefly outshine an entire galaxy of 100 billion stars, which means it can be seen across immense tracts of universe.

Incidentally, the bright light from a supernova is less than 1% of the energy released. 99% is released as neutrinos.

Besides 'core-collapse' supernova, there's a second important type. Occurs in binary system in which one star has evolved to a white dwarf.

Matter sucked down onto white dwarf from companion star triggers runaway nuclear reactions. Star blows itself apart as supernova.

Key significance of second-type supernova – technically, Type Ia – is that explosion is believed to be always same luminosity.

Type Ias have been crucial in measuring distances across the universe. In 1998, they revealed the existence of the mysterious 'dark energy'.

75. What if a supernova went off nearby?

Since a supernova can easily burn as brightly as 10 billion Suns, a supernova going off in our cosmic backyard would be a scary sight.

If a supernova detonated within 30 light years of Earth, it would be a blindingly bright star at least 100 times brighter than Full Moon.

Not only would it be visible daylight but, as faded over next months, would banish night, making it hard for nocturnal creatures to hunt.

But, although light would take just 30 years to reach us, coming behind & taking 300 years, would be deadly sleet of subatomic particles.

When particles struck atmosphere, they would strip Earth of its ozone layer, which shields life against deadly solar ultraviolet light.

Life on Earth's surface would be impossible. The only creatures that would survive would be in sea, caves or underground.

Impossible to estimate how common supernovae are in Milky Way since they are often hidden behind curtains of interstellar dust. But . . .

. . . in galaxies like ours, we see 1 supernova every 50 years or so. Implies 200 million supernovae in 10-bn-year history of Milky Way.

Since birth of Earth, one or two supernovae must have detonated within 30 light years. Could easily have caused a mass extinction of life.

Fortunately, the nearest known supernova in past 400 years – SN1987A – was 170,000 light years away in satellite galaxy of Milky Way.

Bad news: Betelgeuse – prominent star in constellation Orion – on verge of going supernova. Good news: might take another million years!

Luckily, Betelgeuse is about 650 light years away. If it exploded, it would appear ~500 times fainter than supernova at 30 light years.

But a supernova pales into insignificance next to a 'gamma ray burst', unusually energetic supernova which gives birth to black hole.

Crucially, the energy of a gamma ray burst is directed in a single direction like a lighthouse beam. 'Death ray' of high-energy gamma rays.

Gamma ray burst even 10,000 light years away could smash, or 'ionise', atoms in atmosphere, disrupting ozone layer & threatening life.

76. What are neutron stars and pulsars?

Bonkers fact: you could fit the entire human race in the volume of a sugar cube. Why? Because matter is mind-bogglingly empty.

Naively, you can think of an atom as a mini Solar System, with electrons orbiting like planets around a tiny central 'nucleus' like the Sun.

But picture of atom as mini Solar System fails to convey how amazingly empty the atom is. It is 99.9999999999999% empty space.

If you could squeeze all the empty space out of all the atoms in all the people in the world, humanity would indeed fit in a sugar cube.

Not just mad theory. There are objects out in space where all the empty space has been squeezed out of their atoms. Neutron stars.

A neutron star is the relic (imploded core) left when a massive star goes supernova. Imagine Sun squeezed down to the volume of a mountain.

If you could go to a neutron star and scoop out a volume the size of a sugar cube, it would indeed weigh as much as the entire human race.

When a star shrinks to a neutron star, it spins faster like an ice skater pulling in her arms. Spinning neutron stars shout 'Here I am!'

In 1967, Jocelyn Bell, 24-year-old student, was using radio telescope at Cambridge. Found regular pulses of radio waves from object CP1919.

Bell soon found several other pulsing sources. At first, people thought they were ETs signalling us. Dubbed them LGMs, for Little Green Men.

In 1968, Tommy Gold & Franco Pacini realised Bell had found spinning neutron stars. As they spin, they emit radio waves in lighthouse beam.

They were christened 'pulsating neutron stars', or pulsars. Gravity on surface of a neutron star is 100 billion times that on Earth.

So far, three Nobel Prizes have been awarded for pulsars. None has gone to Dame Jocelyn Bell Burnell. Widely considered a major injustice.

77. What are black holes?

A black hole is a region of space where gravity is so strong not even light – fastest thing in universe – can escape. Hence its blackness.

A black hole is believed to result from the death of a very massive star in a cataclysmic explosion known as a supernova.

Paradoxically, when a star blows its outer layers into space, its core implodes, shrinking so fast its density & temperature skyrocket.

If the core is massive enough, no known force can stop the core shrinking to a 'singularity' – a nightmarish point of infinite density.

A black hole consists of a singularity cloaked by an 'event horizon', which marks point of no return for matter falling into the hole.

If the Sun turned into a black hole – don't worry, it's not massive enough for that – its event horizon would be a mere 3 kilometres across.

The gravity of a black hole is so great that it bends light in its vicinity and slows time, according to Einstein's theory of gravity.

So if you could hover close, you'd see the back of your head – light from the back of your head would bend around black hole into your eye.

And thanks to time-slowing, near the event horizon you could watch the future history of the universe flash past like movie in fast-forward.

Black holes are impossible to see directly (so far) because 1) small and 2) black. We infer their existence indirectly from their gravity.

For instance, Cygnus X-1 is a very massive star whirling around invisible companion (black hole). We see X-rays from matter sucked into BH.

Birthcry of black hole should be burst of gravitational waves, vibrating space like ripple on pond. Detection will prove black holes exist.

In addition to 'stellar' black holes, universe contains 'supermassive' ones (in galactic cores), millions to billions times mass of Sun.

There is an outside possibility that universe also contains mini black holes, relics from the violence of the Big Bang fireball itself.

Actually, black holes are not entirely black! As Stephen Hawking discovered, due to 'quantum' effects, they radiate 'Hawking radiation'.

78. Are the stars artificial?

It's a completely daft question – right? Well, actually, it has a bearing on a crucial scientific question: How will we recognise ETs?

The search for extra-terrestrial intelligence (SETI) scans heavens for broadcasts at a single frequency – ET equivalent of radio station.

Such a signal is regular, repetitive – has a pattern. But, information with a pattern contains redundancy. It can be compressed further.

Conclusion: a really efficient signal will have no pattern. Will look random, like the radio emission from the Sun, or an electrical storm.

This is exactly the way our mobile phone transmissions & computer data are now. For efficiency, all repetition/pattern has been removed.

Conclusion: ET signal from advanced civilisation will look random like natural signal. Very hard to distinguish from cosmic radio 'noise'.

And, just as ET signals won't look like our regular signals, ET artefacts won't look anything like our engineered artefacts either.

According to Stephen Wolfram, inventor of the computer language, Mathematica, ET artefacts will look natural like trees . . . and stars.

In all seriousness, Wolfram asks: 'Are the stars artificial?' Although it may appear unlikely, it may be impossible to know for certain.

The Milky Way

79. What does our Milky Way galaxy look like?

The Milky Way is a misty band of white light in the night sky. Appeared to ancients like milk spilled across blackness – hence lyrical name.

In 1610, when Galileo turned his telescope on the heavens, he found the Milky Way was actually made of countless stars, crowded together.

Discovery (1922) that 'spiral nebulae' are islands of stars in ocean of space suggested that Milky Way is one such island, or 'galaxy'.

But it proved very difficult to discern the detailed structure of the Milky Way from the vantage point of the Sun, embedded deep inside.

Visible light from distant stars in the Milky Way is absorbed by curtains of dust hanging across interstellar space.

To see Milky Way's structure, you need a type of light that penetrates the dust. (And even then it's hard to see structure from within.)

Radio waves penetrate dust. They reveal Milky Way is indeed a spiral galaxy. Giant pinwheel of ~200 bn stars, turning ponderously in space.

Milky Way's stars are concentrated in a flattened disc. Seen edge-on, the galaxy has the shape of two fried eggs, back to back.

Like all spiral galaxies, Milky Way has a central, spherical 'bulge' of stars from which jewelled 'spiral arms' of stars snake outwards.

The flattened disc of the Milky Way is about 100,000 light years in diameter. But it is thin – a mere 2000-odd light years top to bottom.

The Sun is located in a 'spur' of the 'Perseus Arm', about 27,000 light years from the Milky Way's centre & about halfway out to the edge.

Sun orbits centre of galaxy about once every 220 million years. Last time it was at its current location, the dinosaurs ruled the Earth.

80. Where are stars born in the Milky Way?

The clue to where stars are born came from Walter Baade, using 2.5-m Mount Wilson telescope during WWII black-outs over Los Angeles.

Baade, a German émigré barred as 'enemy alien' from American war work, discovered Milky Way contains two distinct 'populations' of stars.

Population I, in spiral arms, is dominated by hot blue stars. Population II, in the Milky Way's 'bulge', is dominated by cool red giants.

Crucially, red stars are old whereas blue stars are young. Baade had therefore discovered that the spirals arms are stellar nurseries.

The reason stars are born in spiral arms became clear only when astronomers realised what exactly spiral arms are.

Spiral arms are not permanent features of the Milky Way. If they were, as the galaxy rotated, the arms would inevitably 'wind up'.

Turns out the gaseous disc of the Milky Way is vibrating like surface of pond. Rippling outwards from centre is a 'spiral density wave'.

As density wave moves outward, it compresses interstellar gas in its path, triggering globules to start shrinking to form stars.

It is because the spiral density wave causes a firestorm of star formation as it passes that spiral arms are home to stellar nurseries.

Milky Way's spiral like Galactic Mexican wave! It looks permanent only because our lives are short compared to the time wave takes to pass.

Remarkably, Saturn's rings are spirals similar to spiral arms, just more tightly wound. Both caused by same phenomenon: spiral density wave.

81. What are open and globular star clusters?

Stars are not born alone but in groups of 10s or 1000s, their ferocious heat eating away the gas on the edge of a Giant Molecular Cloud.

Newborn stars gradually disperse since stellar nurseries are violent places, buffeted by fierce stellar winds and stellar explosions.

After a few 100m years, stars of such an unbound, or 'open cluster', may be so separated that it is hard to tell they once were siblings.

In fact, some of Sun's siblings may still be in the solar neighbourhood. Difficult to tell, though, since Sun born 4.55 billion years ago.

Young open clusters, however, are easy to see. Newborn, hot stars in Pleiades cluster (Taurus) are still shrouded in placental nebulosity.

But open clusters, made of stars born together & dispersing, are not only clusters in Milky Way. Also non-dispersing 'globular clusters'.

A globular cluster contains many stars – 100,000 to a few million – compressed into a tight knot only a few tens of light years across.

Globular clusters buzz like bees around the Milky Way's disc. The spiral disc is embedded in a giant spherical swarm of 150–200 of them.

Other galaxies such as the giant elliptical galaxy, M87, don't just have a few hundred globular clusters. They have more than 10,000.

Stars in globular cluster are packed so closely they may even collide, something almost impossible for widely separated stars like the Sun.

For a planet in a globular cluster, the night sky is studded not with 1000s of stars as on Earth but with 100,000s. What a sight it must be.

Globular cluster differs from open cluster not only in being bound rather than unbound but in containing ancient rather than newborn stars.

Clue to origin of globulars comes from ages of stars. Born 10 bn years ago, when spherical gas cloud was still shrinking to form Milky Way.

Ultimately, however, question of how and why globular star clusters formed in the early days of the Milky Way remains a mystery.

82. How many satellite galaxies orbit our Milky Way?

Just as planets have satellites (moons), galaxies have satellite galaxies. The Milky Way has about 25 of them in its gravitational thrall.

Two biggest satellites – Large and Small Magellanic Cloud (LMC & SMC) – are easily visible with the naked eye from the southern hemisphere.

LMC is a faint cloud-like smudge against the night sky – about 10 x apparent size of Moon. SMC is similar smudge about 5 x size of Moon.

Magellanic Clouds named after Ferdinand Magellan, first European to record them – during round-the-world voyage between 1519 and 1521.

LMC: about 10% mass of the Milky Way & about 170,000 light years distant. SMC: about 200,000 light years away, about 5% mass of Milky Way.

In 1987, LMC hosted first naked-eye supernova since 'Kepler's star' of 1604. For about a month, SN1987A pumped out light of 100m Suns.

LMC and SMC are merely largest and brightest of the satellite galaxies that flit about the Milky Way like moths around a candle flame.

Most contain few stars & are super-faint. Biggest about 1000 light yrs across, less than 1% diameter Milky Way; smallest ~150 light years.

Other galaxies also have satellite galaxies. For instance, the Milky Way's giant neighbour, Andromeda, is known to have at least fifteen.

The Milky Way's satellite galaxies pose a major puzzle because there should be about 100 times as many as astronomers observe.

Theory of galaxy origin says dark matter forms clumps (haloes), then drags in normal matter. Key feature: dark matter haloes of all sizes.

A big halo like one that 'seeded' the Milky Way should have gathered about it 1000s of minihaloes – the seeds of small satellite galaxies.

So where are Milky Way's satellite galaxies? Dark matter proponents say they are there but we haven't yet seen them because too faint.

But possibility remains that 'missing satellite galaxy problem' is telling us there is something wrong with the theory of dark matter.

83. What is the major component of the Milky Way?

Milky Way, like all spiral galaxies, is an island of stars & nebulae, right? Wrong. Like an iceberg, most of galaxy is missing from view.

Missing stuff became apparent when people studied stars in outer reaches of Milky Way and measured how fast they are orbiting the centre.

Far from centre of galaxy, stars are moving too fast. Like kids on a speeding-up roundabout, should be flung off into intergalactic space.

Astronomers explained anomaly by postulating galaxy contains invisible, 'dark', matter, whose extra gravity was gripping hold of stars.

Conclusion: flattened spiral of Milky Way is embedded in vast, spherical 'halo' of dark stuff, perhaps 10 x as massive as visible galaxy.

But what is the dark matter? Your guess is as good as anyone's. Current favourite candidate is hitherto undiscovered subatomic particles.

If Milky Way is mostly dark matter, inevitable it is around you now. Many experiments looking for it. Nobel Prize for whoever finds it.

Also evidence from beyond Milky Way that universe contains dark matter. It outweighs the visible matter we are made of by factor 6 or 7.

But dark matter is least daring proposal to explain why outermost stars in Milky Way are orbiting too fast. There is another idea: MOND.

MOND (modified Newtonian dynamics), proposed by Mordehai Milgrom in 1983, can explain speeded-up motion of stars in all spiral galaxies.

Idea: stars gripped not by extra gravity of dark matter but gravity which is stronger in galactic outskirts than predicted by Newton.

MOND embraced by sizeable minority of astronomers. But no one sure of 'deep' physics underlying it. Until known, most physicists sceptical.

The case for dark matter will be bolstered if a candidate particle turns up at the Large Hadron Collider 'atom smasher' near Geneva.

84. What lurks in the heart of the Milky Way?

In the heart of the Milky Way, stars are packed hundreds of times closer together than they are in the neighbourhood of the Sun.

On a planet orbiting a star in the galactic centre, hundreds of thousands of stars would be visible in the night sky.

Centre of galaxy is a place of extreme violence. Great tsunamis of interstellar gas collide with each other, driven by supernova explosions.

Lurking in the dark heart of the Milky Way, 27,000 light years from the Sun and cloaked by choking interstellar dust, is Sagittarius A*.

Sagittarius A* is a black hole 4.3 million times the mass of the Sun, a black widow-like monster devouring gas and ripped-apart stars.

'Event horizon' of Sagittarius A* – point of no return for in-falling matter – is about 15 million km across (1/10 Sun–Earth distance).

No one knows origin of Sagittarius A*. But 'supermassive' black holes – some 1000 times bigger – appear to lurk in heart of most galaxies.

Universe's black holes are either too small (stellar) or, if supermassive, too far away for us to zoom in on with our current telescopes.

Sagittarius A*, which is moderate in size and relatively nearby, is the only black hole we have a realistic chance of imaging.

Very Long Baseline Interferometry (VLBI), which harnesses radio dishes to simulate Earth-sized telescope, is zooming in on Sagittarius A*.

VLBI mere factor of 2–3 away from 'seeing' Sagittarius A*'s event horizon. Should achieve feat in the next few years and prove black hole.

85. What are our closest galactic neighbours?

The Milky Way is one of the biggest members of a mini-cluster of about 30 galaxies known to astronomers as the Local Group.

The only other galaxy of comparable size in the Local Group is the Andromeda galaxy, a giant spiral similar to the Milky Way.

Big spirals like Milky Way/Andromeda are exception in the Local Group. Most others are dwarf galaxies. Biggest only 1/10 as many stars.

Andromeda is most distant object visible to the naked eye. Appears in the sky as an elongated smudge about 6 x the size of the Moon.

At 2.5 million light years, we see Andromeda as it was when the human race's man-ape ancestors were scrabbling about on an African plain.

Andromeda currently falling towards the Milky Way. In 2.3 billion years, it will fly past, its gravity ruffling the stars in our galaxy.

But, like a pendulum overshooting its lowest point, Andromeda will swing back. In 5 billion years' time it will smack into the Milky Way.

Result of collision will be giant elliptical galaxy, dubbed Milkomeda. Sun will be kicked from 27,000 to 52,000 light years from the centre.

The nearest big cluster of galaxies is the Virgo Cluster, about 50 million light years away and consisting of about 1300 galaxies.

In fact, the Local Group is an outlying member of Virgo. It orbits in the quiet suburbs of what astronomers call the 'Local Supercluster'.

Galaxies

86. What are galaxies?

Galaxies – great islands of stars adrift on the ocean of space – are the building blocks of the universe. There are about 100 billion.

The galaxies are flying apart from each other like pieces of cosmic shrapnel in the aftermath of a titanic explosion – the Big Bang.

If the universe were shrunk to a sphere about 1 kilometre across, the 100 billion galaxies would each be about the size of an aspirin.

Some galaxies are regular, some are amorphous splodges of stars. Two most common types are spirals (like our Milky Way) and ellipticals.

Galaxies contain anything from a few million stars – in the case of a dwarf galaxy – to a few trillion – for a giant elliptical galaxy.

Elliptical galaxies are like great bee-swarms of stars. Spherical or slightly elongated. Spiral galaxies, well, they speak for themselves.

Spiral galaxies have a central 'bulge' of old, red stars, and 'spiral arms', where gas is being processed into fresh, new stars.

Ellipticals, unlike spirals, contain hardly any gas. Used up by star formation aeons ago. Consequently, they contain only old, red stars.

Relationship spirals & ellipticals unclear. But it seems ellipticals are created in collisions of two spirals. Star motions randomised.

Some spirals have a curious 'bar' in the centre from which 'spiral arms' extend. There is evidence that our Milky Way is a barred spiral.

Douglas Adams missed a trick. Should have written about the bar at centre of the galaxy rather than restaurant at the end of the universe!

87. How were galaxies discovered?

In the 18th C, astronomers were mad keen on comet hunting. But the night sky contains many misty patches that can be mistaken for comets.

To assist comet hunters, in 1784 Charles Messier catalogues 'vermin of the skies'. Unknown to him, some of these 'nebulae' are galaxies.

In Birr, Ireland, 1845, Lord Rosse builds 72-in telescope – biggest in world. With the 'Leviathan', he finds many nebulae have spiral shape.

Most beautiful spiral nebula is M51. It will later be christened the 'Whirlpool Galaxy'.

Even bigger telescopes, constructed later, show that misty nebulosity of all spiral nebulae is due to innumerable stars smeared together.

Heated debate, 1920. Are spiral nebulae inside our Milky Way? Or are they separate 'island universes' far across ocean of space?

Harlow Shapley maintained spiral nebulae were within the Milky Way; Heber Curtis claimed they were far outside. Dispute settled in 1922.

Edwin Hubble, using 100-in Hooker Telescope on Mount Wilson, near Los Angeles, sees Cepheid variables in Great Nebula in Andromeda.

Period of fluctuation of Cepheid light related to intrinsic luminosity. Hubble deduces Andromeda million or so light years beyond Milky Way.

Andromeda and other spiral nebulae are therefore at enormous distances from Milky Way. Separate islands of billions of stars.

Hubble had discovered fundamental building blocks of universe. 'Galaxies' studded space out to limits probed by the biggest telescopes.

At last the human race knew the true scale of the universe it was lost in. It was unimaginably more vast than anyone had ever dreamed.

88. How do we know how far away galaxies are?

Galaxies are building blocks of universe, so 'How do we know distances of galaxies?' is synonymous with 'How do we know size of universe?'

To find distance of galaxy, necessary to find a 'standard candle' – object whose luminosity we can compare with similar object nearby.

For nearby galaxies, astronomers use Cepheid variables. Period over which they vary their brightness is related to their true luminosity.

Ultra-luminous Cepheids have been spotted in galaxy M100, pinning down its distance as 56 million light years beyond the Milky Way.

Further away, astronomers have to find an even brighter standard candle than Cepheids: Type Ia supernovae.

Type Ia supernovae occur in binaries in which one star dumps matter onto super-compact, Earth-sized 'white dwarf', causing it to explode.

It is widely believed that, when such white dwarfs finally detonate as supernovae, they do so always with exactly the same luminosity.

Type Ia supernovae are so bright they are visible to edge of universe. Have provided estimates of distances of the most remote galaxies.

Cosmic distance measurements allow an estimate of the 'Hubble constant', which sets scale of universe. Best current estimate: 73 km/s/Mpc.

Means that galaxy 1 megaparsec (3.26 million light years) further away than another is receding 73 km/s faster due to Big Bang expansion.

Speed of galaxy revealed by stretching of its light waves (red shift). From this, & knowing Hubble constant, can estimate distance.

Note: distance not terribly meaningful. Because of speed of light, always measuring distance to an object existing 'at an earlier time'.

Rather than distance of a galaxy, astronomers like to refer to it as red shift – a more meaningful measure of its remoteness from us.

89. What are quasars?

Quasars are star-like pinpricks of lights – hence quasi-stellar objects, or quasars – but way beyond distance any star should be visible.

Discovered by Dutch-American astronomer Maarten Schmidt in 1963. Others had seen them but he was first to recognise their significance.

To shine so brilliantly despite being at such staggeringly huge distances across the universe, quasars must be prodigiously luminous.

Typical quasar pumps out 100 x energy of a normal galaxy like Milky Way. Incredibly, it comes from volume smaller than the Solar System.

Nuclear energy woefully inadequate. Only possible source: 'gravitational energy' released by matter falling onto central black hole.

Quasar light emitted by matter heated to incandescence as swirls, like water down a plug hole, through 'accretion disc' onto black hole.

We are not talking about a normal, 'stellar-mass', black hole but a 'supermassive' one. In brightest quasars up to 30 billion x mass Sun.

Long after quasar discovery, people spotted 'fuzz' of surrounding stars. Quasar is super-bright 'nucleus' of galaxy, outshining all else.

Quasars are just extreme examples of 'active galaxies' – whose light is created not principally by stars but by a supermassive black hole.

Active galaxies about 1% of galaxies. In addition to quasars, other types include elliptical 'radio galaxies' & spiral 'Seyfert' galaxies.

It is possible most galaxies – even our own – went through an active (quasar) phase in youth. Ended when central black hole ran out of fuel.

No quasars around today. Their heyday was billions of years ago. Our telescopes show them blazing like brilliant beacons at dawn of time.

90. Do only a few galaxies harbour giant black holes?

For a long time after quasars were discovered, they were thought to be anomalies – cosmic curiosities with no connection to normal galaxies.

Gradually, was realised that most, if not all, galaxies contain in their hearts supermassive black holes – millions to billions x mass Sun.

Most supermassive black holes are quiescent, sitting doing nothing and often hard to see because they are cloaked in interstellar dust.

Even the Milky Way harbours a supermassive black hole, albeit a modest one. Sagittarius A* weighs in at about 4.3 million solar masses.

Strong suspicion most galaxies, including our own, went through a violent quasar phase in youth. Switched off when gas supply ran out.

Quasars may have flared up in early universe because there was plenty of foodstuff around. Gas was then mopped up by star formation.

Also, galaxies closer together back then (universe is expanding). Galactic collisions may have supplied central black holes with fodder.

Supermassive black holes tiny and galaxies big. Yet, oddly, properties are connected. Black holes 1/700 mass of central 'bulge' of stars.

Hints at intimate connection between black holes & galaxies. Either galaxies spawned BHs or BHs spawned galaxies. Or two born together.

The precise nature of the connection between supermassive black holes and galaxies is one of the big unsolved puzzles in cosmology.

91. Why are there giant black holes in galaxies?

In the standard picture, galaxies formed first. Later came the giant black holes, found in the heart of most, if not all, galaxies.

Scenario: Hearts of first galaxies – smaller than today's – crowded with stars. They exploded and left black holes, which collided/merged.

In crowded early universe, galaxies collided, making bigger galaxies. In process, central black holes merged, making them even larger.

Over past 10 billion years, such 'supermassive' black holes (SMBHs) have continued to grow, gorging on gas & stars surrounding galaxies.

Problem with scenario: chaotic. Doesn't explain why central holes invariably are 1/700 mass of host galaxy's central 'bulge' of stars.

Possible explanation: 'jets'. Supermassive black holes are spinning & often launch thread-thin jets of matter from their poles (spin axis).

Jets thought to be driven by pent-up energy of magnetic fields, viciously twisted in super-hot 'accretion' disc swirling around black hole.

When jets form, they blow away gas – raw material for new stars – snuffing out star formation. May explain relation SMBH mass & bulge mass.

But some, including Joe Silk at Oxford, think we've got the scenario back to front. Galaxies don't create SMBHs; SMBHs create galaxies.

In Silk's view, after Big Bang, cooling debris congealed into giant gas clouds – precursors of galaxies – but sat there, doing nothing.

Cores of some were so dense they shrank under self-gravity & formed SMBHs. Jets switched on, stabbing millions of light years across space.

Where jets slammed into an inert gas cloud, they compressed the gas, triggering a rash of star formation – creating a new galaxy.

Evidence: quasar HE0450-2958, floating in space about 23,000 light years from a galaxy (about distance of Sun from centre Milky Way).

HE0450-2958 appears to have no surrounding galaxy. Only known 'naked quasar' – supermassive black hole floating isolated in the void.

Crucially, the jet from naked quasar stabs like a laser beam into the galaxy. Some believe the quasar jet gave birth to the galaxy.

92. How did giant black holes get so big so fast?

Some of the most distant quasars, which formed shortly after Big Bang, already contained black holes of 10 billion times mass of the Sun.

The existence of monster black holes so early in the universe's history poses a difficult question: How did they get so big so quickly?

In standard picture, stellar-mass black holes in first galaxies merged to form bigger holes. When galaxies collided, their holes merged.

Such a multi-step process must have proceeded very fast indeed to explain monster black holes already present in most distant quasars.

American astrophysicist Mitchell Begelman has suggested an alternative, and quicker, way supermassive black holes may have been created.

Deep inside cloud shrinking to form one of first galaxies, gas gets so dense that giant black hole forms – without first forming stars.

Central black hole grows rapidly, feeding on surrounding gas, which is delivered to it at fast rate since gas cloud is still shrinking.

The scenario is of a giant glowing ball of gas – which Begelman calls a 'quasistar' – with a black hole growing hidden away inside it.

Usually, if BH gets big, its heat blows surrounding gas away, limiting its growth. But Begelman's black hole can grow at prodigious rate.

Like a parasitic wasp inside a caterpillar host, says Begelman, the black hole gradually eats its way out of its cocoon.

Eventually, it blows away the last shreds of surrounding gas. Hey presto, there emerges . . . a fully formed giant black hole.

Proving the scenario will be a challenge. However, quasistars will pump out prodigious quantities of heat (infrared light).

It is conceivable that they will be detected by the James Webb Telescope, which will image in the infrared & is due for launch in 2018.

93. What are the largest structures in the universe?

The universe's 100 billion galaxies are not dotted evenly around space. Instead, they huddle together in clumps, or 'clusters'.

But, just as galaxies crowd together into clusters, galaxy clusters themselves congregate in even bigger 'superclusters'.

The Milky Way belongs to collection of ~30 galaxies called the Local Group. It is attached to a local supercluster called the Virgo Cluster.

But even superclusters are not sprinkled evenly around the universe. They too huddle together into even bigger agglomerations.

In places, great daisy chains of superclusters snake across cosmos; in others superclusters form curtains, or 'walls', across space.

The Sloan Great Wall has mass of about 10,000 normal galaxies & stretches 1.4 billion light years (1/60 diameter observable universe).

The Sloan Great Wall even earned a place in the 2006 *Guinness Book of Records* as the 'largest structure in the universe'.

Existence large structures early in cosmic history possibly poses a problem to astronomy. How could they have formed so soon after Big Bang?

Largest structures are remnants of random 'quantum fluctuations' in energy in first split-second of universe, magnified to enormous size.

Surprisingly, the biggest galaxy groupings in today's universe were seeded by primordial structures in Big Bang smaller than a single atom.

94. Is it possible the galaxies we see are an illusion?

In 1977, the MERLIN array of radio telescopes, centred at Jodrell Bank in Britain, imaged two quasars that seemed remarkably similar.

The two quasars of QSO 0957+561 were more than similar – they were the same. 'Double quasar' was first 'gravitationally lensed' object.

Gravitational lensing is a version of light bending by gravity predicted by Einstein's theory of gravity (general relativity) in 1915.

If a massive object (eg: galaxy cluster) is between us & a distant object (eg: quasar), its gravity can bend/focus light of distant object.

Since multiple paths possible for light around intervening object, multiple images. Most possible is 5. Some too faint to be easily seen.

G-lensing by intervening object (lens) not only focuses light of distant object but concentrates it, boosting, or magnifying, brightness.

Therefore gravitational lensing acts like nature's telescope, boosting the brightness of objects ordinarily too far away to be seen.

Nearby galaxies that we see are really there. But further away, more chance of intervening lenses. Distant universe is more illusory.

BTW, gravitational lensing does not always produce neat images. The gravity of a cluster of galaxies distorts distant object into arcs.

Gravitational lensing can be turned on its head to reveal distribution of intervening stuff doing lensing, even if too dark to see directly.

Large Synoptic Survey Telescope (Chile) will use weak lensing to deduce distribution of matter – particularly dark matter – in universe.

Telescopes were invented to image light. Ironically, the Large Synoptic Survey Telescope will image darkness!

Claudio Maccone suggested using Sun as gravitational lens. 'Gravity telescope' would have focus way beyond Pluto, so challenge to build.

95. Why are telescopes considered time machines?

Light, though fast, is not infinitely fast. It takes time to reach us from objects. So, we see things as they were at earlier times.

Delay effect is unnoticeable for everyday objects, since speed of light is enormous – 300,000 km/s (million x faster than passenger jet).

But universe is BIG & distances huge. Light takes long time to travel to us from astronomical bodies. Telescopes in effect 'time machines'.

We see Moon as it was 1.3 sec ago; Sun 8.3 min; nearest star 4.2 years; most distant naked-eye object (Andromeda galaxy) 2.5m years.

Impossible to know what any astronomical object looks like 'now' (meaningless concept), only what it looked like when light left it.

Say, light so slow it takes 100 years to cross street. House visible on far side could have gone long ago. Same for distant galaxies.

Most distant galaxies probably don't exist any more. We see them as they were more than 10 billion years ago – long before Earth was born.

Furthest back we can see with light (actually, form known as radio waves) is 13.7-billion years, 380,000 years after birth of universe.

13.7-bn-year-old light is 'afterglow' of Big Bang. Before then, universe was filled with 'fog'. Light unable to travel in straight line.

1% of static on TV tuned between stations is 'afterglow' of Big Bang – very first light. Accounts for 99.9% of all photons in universe.

The Universe

96. How big is the universe?

In order to answer How big is the universe? – it is first necessary to define what exactly is meant by the 'universe'.

Key fact: the universe has not existed forever. It was born. It exploded into being 13.7 billion years ago in titanic explosion – Big Bang.

Fact: Universe born means we see galaxies whose light has taken less 13.7 bn years to get here. More distant objects – light still on way.

The objects we see – about 100 billion galaxies – are in a giant 'bubble' of space, centred on Earth, known as the 'observable universe'.

Distance to edge of observable universe is about 42 billion light years, making observable universe about 84 billion light years across.

Q: How can edge be 42 bn light yrs away if universe only 13.7 bn yrs old? A: Early on, universe expanded, or 'inflated', faster than light!

Note: Light speed maximum speed only in Einstein's special relativity (1905). In general relativity (1915), space can expand at any rate.

Observable universe is bounded by imaginary boundary called the 'light horizon', which marks furthest possible seeable with a telescope.

But 'cosmic horizon' is much like horizon at sea. Just as we know more ocean over horizon, we know more of universe over cosmic horizon.

According to the theory of 'inflation', effectively there may be an infinite amount of the universe beyond horizon. Universe infinite!

97. What was the Big Bang?

About 13.7 billion years ago, all space, time, energy and matter erupted into being in a titanic fireball called the Big Bang.

Ever since the Big Bang, the universe has been expanding, the galaxies – like our own Milky Way – congealing out of the cooling debris.

History. In 1916, Albert Einstein applies his theory of gravity (general relativity) to the biggest gravitating mass – the universe.

His theory tells him that the universe must be in motion. But Einstein, believing it is unchanging, misses the message in his own equations.

Aleksandr Friedman (1922) & Georges Lemaître (1927) independently recognise truth. Universe is not 'static' but 'evolving' (from Big Bang).

Lemaître, Belgian Catholic priest, sees parallel between Universe born in Big Bang fireball & Bible's 'let there be light' story of Genesis.

In 1929, Edwin Hubble discovers that universe is expanding. All but closest galaxies are fleeing from us. The further away the faster.

If galaxies are flying apart, they were closer together in past. 13.7 billion years ago they were on top of each other. That was Big Bang.

When universe was smaller, it was also hotter (just as air compressed by bicycle pump gets hotter). Therefore Big Bang was a 'hot' Big Bang.

In 1948, Hoyle, Bondi & Gold, propose 'steady state theory'. The Universe is expanding but new matter born in gaps, so Universe always same.

Ironically, term 'Big Bang' is coined by Fred Hoyle – who never believes in the Big Bang theory – during a BBC radio programme in 1949.

Early 1960s, radio telescopes reveal quasars in distant universe. Non-existent in today's universe. Evolving universe supports Big Bang.

1965, Arno Penzias & Robert Wilson find heat 'afterglow' of Big Bang fireball – cosmic background radiation. Big Bang theory triumphant.

98. Where did the Big Bang happen?

The term 'Big Bang' conveys the wrong visual picture in almost every conceivable way. In particular, it gives impression of an explosion.

An explosion, like that of dynamite, happens at one place. But there is no place you can point to & say 'The Big Bang happened there'.

In the Big Bang, space burst into being and immediately began growing everywhere. It happened everywhere at once.

Imagine raisins in a rising cake. From point of view of any raisin, all other raisins recede (think of an infinite cake with no edge!).

Galaxies embedded in expanding space are like raisins in rising cake. From point of view of any galaxy, all other galaxies receding.

So, in an expanding universe, everyone sees the same view. Everyone appears to be at the centre of the explosion (though nobody is).

Also, in an explosion of dynamite, shrapnel explodes in pre-existing space. But for the universe there was no pre-existing space.

The Big Bang did not expand into anything. Space simply appeared and every bit of it began expanding away from every other.

Think of infinite cake again. If it is infinite, there is no outside for it to expand into. Expansion means all points inside grow apart.

Of course, it could be space curves back on itself like high-dimensional version of balloon surface. Again no 'outside' to expand into.

If your brain hurts, remember this: Big Bang 4-dimensional (3 space, 1 time) & so fundamentally unvisualisable to 3D beings like us.

All we can do is catch glimpses of Big Bang, can never grasp in its entirety. Only mathematical theory of general relativity can do that.

99. How do we know there was a Big Bang?

Universe expanding, so must have been smaller in past. Universe's helium (10% of atoms) can be explained only if made furnace of Big Bang.

Everyday evidence: 1% of the static, or 'snow', on a TV tuned between the stations has come directly from the Big Bang.

Fireball of Big Bang like fireball of H-bomb. But, whereas heat of bomb dissipates into surroundings, not possible for Big Bang.

The heat of the Big Bang had nowhere to go. It was bottled up in the universe – which, by definition, is all there is.

Heat of Big Bang still around, greatly cooled by universe's expansion in past 13.7 bn years. Now only 3° above absolute zero (–270 °C).

Instead of appearing as visible light, 'afterglow' of Big Bang appears as microwaves (& mm waves) – invisible light picked up by your TV.

Before striking your TV aerial, Big Bang microwaves had been travelling for 13.7 bn years & last thing they touched was Big Bang fireball.

A whopping 99.9% of all photons (particles of light) in universe are not from stars and galaxies but from the afterglow of the Big Bang.

If we could see universe from outside, 'afterglow of creation' would be most striking feature. All space glowing like inside of a light bulb.

The atmosphere and all cool objects (even you) glow with microwaves, so ironically the dominant light in the universe is difficult to see.

'Cosmic background radiation' discovered only in 1965, and then by accident by two radio astronomers at Bell Labs, Holmdel, New Jersey.

Despite thinking they've picked up microwave glow of pigeon droppings (birds nesting in radio horn), Arno Penzias & Robert Wilson get Nobel.

Cosmic background radiation carries with it a priceless 'baby photo' of the universe when it was a mere 380,000 years old.

Locations where afterglow is warmer/colder than average reveal first clotting of matter after Big Bang – 'seeds' of galaxies.

For their part in finding 'seeds' of the first cosmic structures, John Mather and George Smoot also received 2006 Nobel Prize in Physics.

100. What happened before the Big Bang?

In the beginning was the 'false vacuum', goes the standard story. It had an unusual property – repulsive gravity – so it 'inflated'.

The more vacuum that was created, the more the repulsive gravity, and the faster the vacuum expanded. Faster and faster.

The more vacuum was created, the more energy there was too. Energy from nothing – another unusual property. The 'ultimate free lunch'.

But the false vacuum was unstable. Bits 'decayed' randomly to 'true' vacuum – our vacuum. Imagine bubbles forming in a vast ocean.

Energy of false vacuum had to go somewhere. Created matter in bubble-universes & heated it to tremendous temperature. Made hot Big Bangs!

In this 'inflationary' picture, our universe is but one among a vast number, forever separated by ever-growing expanses of false vacuum.

When inflation ran out of steam, normal expansion began. Likened to explosion of stick of dynamite compared to H-bomb of inflation.

Where did high-energy false vacuum come from? Quantum theory permits energy to pop into existence from nothing (Heisenberg uncertainty).

Perhaps small patch of false vacuum popped into existence/ began expanding. Inflation unstoppable since vacuum grows faster than eaten away.

Obvious next question: Where did the laws of physics (quantum theory) that allow spontaneous creation of energy from nothing come from?

Infinite regress. Maybe no better than saying universe rests on back of turtle. Which prompts question: What is the turtle on the back of?

As a lady attending a cosmology lecture by Bertrand Russell said: 'You're very clever, young man. But it's turtles all the way down!'

101. How fast is the universe expanding?

Universe's expansion rate quantified by Hubble constant. Best current estimate: 73 km/s/Mpc (1 Megaparsec = 3.26 million light years).

This means a galaxy 3.26 million light years further away than another is receding 73 km/s faster due to Big Bang expansion.

However, the universe has not always expanded at the rate it is expanding today. In fact, its expansion rate has had a chequered history.

Naively, might think universe began expanding fast from Big Bang & has been slowing since, as expansion runs out steam. But more complex.

Initially, only vacuum. This 'inflated' at phenomenal speed, doubling size at least 60 times over in first split-second of universe.

When 'inflation' over, tremendous vacuum energy dumped into creating matter & heating it to enormous temperature. This was the hot Big Bang.

After inflation, universe expanded at far more sedate rate, gradually slowing because of braking effect of galaxies pulling on each other.

But recently – in the past few billion years – big surprise. The universe's expansion, which has been slowing, has speeded up again.

Astronomers believe empty space contains weird kind of energy. Repulsive gravity of this 'dark energy' accelerates expansion of universe.

Obvious question: Is there connection between accelerated expansion of inflation & accelerated expansion of dark energy? No one knows.

If dark-energy-driven expansion continues, will push galaxies away. By AD 100 billion, Milky Way will be only galaxy in observable universe.

102. Why is the sky dark at night?

First person to ask this question, in 1610, was Johann Kepler, imperial mathematician to the emperor of the Holy Roman Empire.

In Padua, Galileo's newfangled telescope had revealed stars invisible to naked eye. Kepler wondered: What if stars march on forever?

Just as if you look into a dense pine forest you see nothing but trees, if you look out into the Universe, should see nothing but stars.

Kepler concluded: contrary to expectations, night sky should not be black but as bright as surface of typical star. As bright as the Sun!

Actually, typical star is a 'red dwarf' – they account for 70% of all stars – so night sky should be blood red from horizon to horizon.

Puzzle of why night sky is dark rather than bright became known as Olbers' Paradox after 19th-C German astronomer who popularised it.

Edgar Allan Poe had most plausible explanation. Perhaps sky is dark because light from the most distant stars hasn't arrived at Earth yet.

Poe's idea supported by discovery of finite age of universe. We see only objects whose light has taken less than 13.7 bn years to get here.

But Big Bang does not explain paradox because – well, there is no paradox! Even in infinitely old universe, night sky would not be bright.

Kepler made tacit assumption stars can burn forever and can pump out an unlimited amount of light into space. This is incorrect.

In fact, even if all stars in universe turned all their fuel into starlight, not enough to fill space and make Earth's skies glow at night.

Think of bath with too little water to fill it. The universe is the same. The stars simply contain too little energy to fill it with light.

Who would have thought that the darkness of the night sky would have been a puzzle for 400 years and triggered so much cosmic thinking!

103. What is dark matter?

No one knows. Unlike normal matter (atoms), it gives out no light – or too little light to be detected by our best current instruments.

It outweighs the universe's visible matter – the stars and galaxies – by a factor of between six and seven.

We know of its existence because its gravity tugs on the visible stars, making them move as if more matter is present than we can see.

First evidence 'missing mass': vertical motions of stars in Milky Way disc too fast. In 1932, Jan Oort claimed reacting to invisible matter.

Then, in 1934, Fritz Zwicky found that galaxies in clusters orbiting cluster centre too fast. Reacting to gravity of invisible material.

In 1980s, Vera Rubin found stars in outskirts of spiral galaxies orbiting too fast. Like kids on sped-up roundabout, should be flung away.

Stars do not fly off into intergalactic space, say astronomers, because they are gripped by the gravity of invisible, 'dark', matter.

Every spiral galaxy – including our Milky Way – believed to be embedded in giant spherical 'halo' of dark matter, far outweighing stars.

Identity of dark matter is one of biggest puzzles in physics. Candidate favoured by most scientists: hitherto undetected subatomic particle.

Since dark matter is ubiquitous, it must be passing through the Earth. Several experiments have been set up in mines to find it.

There is realistic possibility that dark matter candidate will be found by Large Hadron Collider – the giant 'atom smasher' near Geneva.

Whoever solves the puzzle of the dark matter will undoubtedly earn themselves a Nobel Prize.

104. What is dark energy?

It is invisible. It fills all of space. And it has repulsive gravity, which is speeding up the expansion of the universe.

Dark energy discovered in 1998 by 2 teams – one led by American Saul Perlmutter and other by Australians Nick Suntzeff & Brian Schmidt.

'Standard candle' supernovae in remote universe fainter than expected. Must be driven further away by more than normal cosmic expansion.

Contrary to all expectations, cosmic expansion is speeding up (should be slowing because of 'braking' effect of gravity between galaxies).

Space itself must contain some kind of 'springy' stuff – dark energy – countering the gravity trying to pull all the galaxies together.

Dark energy is the major component of the universe. It accounts for 73% of all cosmic mass-energy (dark matter 23%; ordinary matter 4%).

Incredible to find universe's major constituent only in 1998. Lesson to physicists, back to 19th C, who've claimed little more physics left.

Actually, dark energy is very dilute. But effect accumulates. Explains why unnoticeable on Earth whereas it holds sway over cosmic volumes.

Without doubt, dark energy is one of the most unexpected discoveries in the history of science. Also, one of the most baffling.

Quantum theory has given us computers & lasers & nuclear reactors; understanding of why Sun shines & why ground beneath our feet is solid . . .

. . . yet when quantum theory is used to predict energy of vacuum – dark energy – get number 1 followed by 120 zeroes bigger than we observe.

Dark energy represents the biggest discrepancy between a prediction and an observation in the history of science. Something wrong?!

Most physicists believe that some 'big idea' is missing. And, only when someone finds it, will we finally understand dark energy.

105. Is the universe tailor-made for life?

It does appear to be, though we should be very cautious on the matter, because appearances in science can be deceptive.

If gravity few % stronger, it would squeeze/heat Sun's core, so fuel used up in 1 bn years – not enough for evolution of intelligent life.

On other hand, if gravity few % weaker, Sun's core would not be squeezed/heated enough to burn at all. Life could not have arisen on Earth.

Similarly, if nuclear force few % stronger, the Sun, rather than burning fuel in 10 bn years, would burn in less than 1 second & explode!

Everywhere we look in nature, it seems laws of physics are 'fine-tuned' for us to be here. Question: What are we to make of this?

One possibility – though not scientific – God fine-tuned laws of physics. However, no evidence for supernatural influence running universe.

Other possibility – there are many universes, each with different laws & we find ourselves in one fine-tuned for life. How could we not?!

Topsy-turvy idea that laws of physics are the way they are because otherwise we would not exist to notice is called 'anthropic principle'.

Caution: As yet, there's no deep 'theory of everything'. Might show nature's forces interrelated. Might mean less fine-tuning than we think.

However, on one matter – the incredibly tiny magnitude of the dark energy – an anthropic explanation appears unavoidable.

Repulsive dark energy must be extremely small to have not hindered shrinkage of gas clouds to form the galaxies essential for our existence.

106. Is there more than one universe?

Nature appears to be banging us over the head and yelling at us that this is not the only universe. Evidence coming from many directions.

Many different versions of the 'multiverse'. Not yet clear how they fit together into a seamless whole. This an emerging paradigm.

We certainly know more of universe over the cosmic 'horizon'. According to theory of 'inflation', infinite number of domains like our own.

Each domain would have had Big Bang. But, out of cooling debris, different galaxies/stars would have congealed. So different histories.

The apparent fine-tuning of laws of physics for us to be here hints also that there are other universes with different laws of physics.

A framework which provides many domains with different laws is 'string theory', in which particles are vibrating 'strings' of mass-energy.

String theory indicates number of universes may be 1 followed by 500 zeroes. (Problem: why are we in this one and not another?)

Since string theory says universe has 10 dimensions, scope for universes not only with different laws but different numbers of dimensions.

Quantum theory also suggests that either atoms exist in many parallel realities, or behave as if they do (most physicists believe latter).

Strong hint of connection between quantum theory's 'many worlds' & alternate histories played out in domains beyond universe's horizon.

Physicist Max Tegmark even believes there may not be one multiverse but a whole hierarchy stacked like Russian dolls.

Life in the Universe

107. How did life get started?

Defining life is difficult, but this may come close: Life is a self-sustained chemical system capable of undergoing Darwinian evolution.

There's no doubt life can arise in the universe. Look in a mirror. At Big Bang, universe was lifeless; now, at least, it contains us.

Universe began with atoms of hydrogen (simplest) & helium (too antisocial to join other atoms). Insufficient to build complex biomolecules.

Nuclear fusion in stars built up atoms of heavier elements, notably carbon, oxygen & nitrogen. Assembly of complex 'hydrocarbons' possible.

Such 'organic' molecules, among which are probably amino acids, are found everywhere in interstellar space. They are the building blocks of life.

In pools of water, under a sheltering sky, first self-replicating molecules formed on newborn Earth. Precisely how this happened is unknown.

First (probably) came simple RNA (ribonucleic acid); only later complex DNA (deoxyribonucleic acid). First replicating 'cells' came later.

Over time, populations of organisms changed as those with optimum traits for survival left most offspring (evolution by natural selection).

Life on Earth arose quickly, almost as soon as newborn Earth was cool enough. Implies non-life to life is easy step (yet impossible in lab).

Needed: molecular building blocks of life, energy to drive reactions between them, solvent such as water in which reactions can occur . . .

Young Earth, bombarded with comets laden with molecular building blocks of life such as amino acids, was apparently perfect environment.

Could life exist without water? Maybe. However, water is most common liquid in universe. Water's unique properties make it hard to replace.

Is life always based on carbon? Maybe not. But carbon is 4th most abundant element, 7 x more than silicon, which also has complex chemistry.

108. Could life exist elsewhere in the Solar System?

Space is harsh. Vacuum, extreme cold and heat, lethal UV radiation & high-energy particles – all extremely damaging to living cells.

If too hot, complex molecules break apart; if too cold, chemistry of metabolism is too slow. Also, need shelter against particles/radiation.

Airless worlds like the Moon and Mercury are almost certainly lifeless. Same is true for most worlds in frozen outer parts of Solar System.

In remote past, Mars was much more like Earth, with denser atmosphere, higher temperatures, and oceans. Life could have originated on Mars.

Martian microorganisms might still survive today in underground pockets of ice or water, shielded from harsh conditions on surface.

Answer should come from future mission to collect Martian sample. Finding a second biology – on Mars – would be scientifically momentous.

At certain depths in thick atmospheres of Venus & Jupiter, bacteria might survive. However, hard to see how life could have got started.

Jupiter's moon Europa has ice-covered ocean, energy source (tides from planet), biomolecules from comets. Even complex life could exist.

Same may be true for Jovian moon Ganymede and for Saturnian moons Enceladus and Titan. Will be hard and expensive to prove true or false.

Find of 'extremophiles' – bacteria in rock, darkness, super-heated water etc – suggests life may thrive at many places in the Solar System.

However, nothing has been found yet. Sobering thought: In all of the universe, Earth is still the only place known to harbour life.

109. Could life have come from space?

Not unlikely. Take nearby Mars. Smaller than Earth. So, after its birth, it would have cooled from molten state more quickly than us.

Evidence on Martian surface of dried-up oceans & rivers. In its first 500m years, Mars would have been Eden. Life may have arisen there.

Add to this fact that we find meteorites from Mars on Earth, knocked off the Red Planet by big impacts and later intercepted by Earth.

So there is a possibility that the Earth was 'seeded' by microorganisms carried inside Martian meteorites. We could all be Martians!

Idea of transport of life between worlds – 'planetary panspermia' – is mainstream. But idea of transport between stars is controversial.

Chandra Wickramasinghe & late Fred Hoyle noticed light of distant stars absorbed by clouds. Claimed absorption pattern matched bacteria.

Wikramasinghe & Hoyle claimed, controversially, that gas clouds floating between the stars are the graveyards of countless dead bacteria.

When stars and planets condense from such clouds, bacteria survive in temporarily melted cores of comets. A few revive and proliferate.

When a comet gets nudged sunward, it can transport bacteria to planetary surfaces like Earth's. Forget Martians, we could be star children!

'Interstellar panspermia' would explain how life got started so quickly on Earth though so far impossible to create from non-life in lab.

If Wickramasinghe & Hoyle are right, life is a cosmic phenomenon. Everywhere in the galaxy we will find DNA-based life like ours.

Even more extreme twist: in 1970s, Francis Crick & Leslie Orgel suggested life on Earth (and around galaxy) deliberately 'seeded' by ETs.

110. Is our Solar System unique?

Solar System has ordered structure: planets all orbit in same direction, more or less in same plane. Probably related to system's origin.

This led philosopher Immanuel Kant (1724–1804) and astronomer Pierre-Simon Laplace (1749–1827) to propose 'nebular hypothesis'.

Idea: planets condensed from flat disc of matter swirling around newborn Sun. Big question: has this happened around other stars too?

In 1980s, Dutch/American infrared satellite IRAS found stars with excess amount of heat (infrared), probably from surrounding dust discs.

In case of star Beta Pictoris, 63 light years away, disc was later imaged with telescope on ground. Probably contains dust grains/pebbles.

In early 1990s, Hubble Space Telescope detected protoplanetary discs surrounding nascent stars in Orion nebula. Discs turn out to be common.

Dusty protoplanetary discs are easier to detect than fully fledged planets: they scatter a star's light and also glow with infrared (heat).

Very young stars may be surrounded by real protoplanetary discs. Older stars may have debris discs, from collisions between larger bodies.

Some discs are truncated, or have empty centres, or gaps, probably caused by gravity of larger bodies, like the gaps in the rings of Saturn.

Computer simulations suggest that gas and dust in a flat, rotating disc is likely to agglomerate into larger bodies, ultimately forming planets.

So all evidence points in one direction: our Solar System is not unique, although other planetary systems may be less orderly than ours.

111. What is an exoplanet?

'Exo' means 'outside', ie, outside our Solar System. Planets in our Solar System orbit the Sun whereas exoplanets orbit other stars.

Number of planets in our own Solar System: 8. Number of known and confirmed exoplanets (spring 2011): over 500.

Exoplanets are hard to see. They're small, dark, & close to their parent star. They reflect only a very small portion of the star's light.

So direct detection is all but impossible. Instead, most exoplanets are found through their indirect influence on their parent stars.

Gravity of orbiting exoplanet causes star to wobble. Hard to see on the sky (astrometry), but quite easy to measure in stellar light.

Star alternately moves towards us and away from us. Result: slight periodic shift in wavelength of stellar light (Doppler effect).

This reveals orbital period and elongation of orbit, and (if stellar mass is known) lower limit for mass of planet.

First exoplanet orbiting Sun-like star was found this way in 1995: 51 Pegasi b, detected by Swiss team led by Michel Mayor.

Other method: if we see orbit edge-on, planet regularly crosses face of parent star. During these 'transits', star is slightly dimmed.

If star's size is known, brightness dip reveals size of planet. Combined with mass (Doppler method) gives density.

To see edge-on orbits, it's necessary to monitor many stars. Being done by NASA's Kepler space telescope. Over 1200 candidates found so far.

Holy grail: Earth-like exoplanet in Earth-like orbit around Sun-like star. Might contain life. Kepler may find it within a few years.

112. What are the strangest exoplanets found so far?

Almost all exoplanets found to date are strange in one way or another. Exoplanets exhibit a bewildering variety of properties.

First exoplanets found (1992) orbited a pulsar (stellar corpse). Origin unclear. No life possible because of pulsar's powerful X-rays.

First planets found around Sun-like stars were 'hot Jupiters' – more massive than Jupiter but closer to their stars than Mercury is to Sun.

Granted, hot Jupiters are easier to spot: massive planets in tight orbits produce bigger stellar wobble. But hundreds of them are now known.

Hottest exoplanet: WASP-12b, 2240 °C. May be slowly evaporating like giant comet. Evaporation observed with other planet, HD 209458b.

Smallest orbit: GJ 1214b (2.1m km). Shortest period: 55 Cancri e (17h 40m). Some planets have highly inclined or very elongated orbits.

Many stars have systems of two or more planets. 55 Cancri has five planets; Gliese 581 and Kepler-11 both have six.

Kepler space telescope even found a candidate system with two planets sharing the same orbit, with one travelling 60° behind the other.

Some rocky planets are so hot they must have a molten surface. Others may be completely covered by an ocean and a hot, steamy atmosphere.

CoRoT-7b and Kepler-10b are rocky planets, a bit larger and heftier than Earth. Both orbit close to their star, so they're lava planets.

Gliese 581g is bigger. It also has a small orbit, but its parent star is a cool red dwarf, so there may be lakes or seas on its surface.

Theorists think some exoplanets could consist mainly of carbon compounds, or iron/nickel, while others could be rocky without a metal core.

Extra-solar systems are very different from Solar System. A planet like Earth, with oceans and life, orbiting Sun-like star, might be rare.

113. Is there any way we can communicate with alien civilisations?

In 19th century, scientists proposed communicating with Martians by planting trees in geometric shapes, or lighting big fires in Sahara.

In 1959 *Nature* article, Giuseppe Cocconi and Philip Morrison suggested 21-cm radio waves is best choice for interstellar communication.

One year later, Frank Drake started Project Ozma. He tuned in to stars Epsilon Eridani and Tau Ceti to search for artificial radio signals.

Since 1960, the search for extra-terrestrial intelligence, or SETI, has become ever more sensitive. But still no ET signal detected.

We've also sent messages on spacecraft – *Pioneer* Plaque and *Voyager* Interstellar Record – and coded radio messages to other stars.

Also, radio & TV broadcasts have turned Earth into a 'natural' emitter of strong artificial radio waves. These could be picked up by aliens.

Through the SETI@Home project, you can participate in the search. Astronomers have also searched at visible wavelengths (optical SETI).

Communication as we know it will be impossible: even in case of nearest star, there will be 8-year delay between question and answer.

Language is also an issue. Mathematicians have devised 'universal languages' that might be intelligible to aliens, if they try hard enough.

SETI's chance of success depends on number of Earth-like planets; frequency of life/intelligence; etc. Factors embodied in 'Drake equation'.

'Eerie silence' (term used by Paul Davies) may suggest that alien civilisations are rare or even non-existent. We might be an aberration.

However, SETI scientists are dogged – if you don't look, you certainly won't find anything. And 50 years is just a blink of a cosmic eye.

114. Have we ever been visited by aliens?

In *2001: A Space Odyssey*, aliens have left a 'baby alarm' buried on Moon to warn them if human race survives & ventures across space.

If space-faring ETs have ever arisen anywhere in our Milky Way, there is a strong argument that they must have visited our Solar System.

Argument due to Enrico Fermi, Italian-American physicist who built first nuclear reactor on abandoned squash court, U of Chicago, 1942.

Fermi said easiest way to explore galaxy is with 'self-reproducing probes'. One flies to nearest star. Uses resources to build two copies . . .

Such space probes could 'infect' the galaxy like a virus. It would take only tens of millions of years to visit all stars in the Milky Way.

Visiting all the Milky Way's planetary systems would therefore take only about 0.1% of the 10 bn years our galaxy has been in existence.

'Fermi Paradox': if aliens exist in our Milky Way, they must have come our way. So, in the immortal words of Enrico: 'Where are they?'

Some says no ETs because we're first civilisation to arise. We are condemned to cosmic loneliness, never to find anyone else to talk to.

Other possibilities include: killer ETs wipe out nascent space-faring civilisations like us OR we're in 'nursery zone', off-limits to ETs.

But absence of evidence not necessarily evidence of absence. If ETs visited Earth, evidence could have been erased by weather/geology.

Most likely place to find ET artefact is on dead world like Moon where it would survive for aeons – exactly like 'monolith' in *2001*.

But 200 billion billion billion km^3 of space within orbit of outermost planet. Have not yet explored enough to say aliens not been here.

The History of Astronomy

115. Who were the first astronomers?

Astronomy is the oldest science. Or so say astronomers. First astronomers – prehistoric people, wondering what the Sun, Moon and stars were.

Daily motion of Sun provided clock. Monthly phases of Moon and yearly rhythm of seasons provided calendar. Stars provided orientation.

Carved animal bone (France, 30,000 BC) may have been first lunar calendar. Cave paintings in Lascaux (15,300 BC) may depict constellations.

Stonehenge, built 3100–1600 BC, was primitive observatory to keep track of seasons. Sun still rises over Heel Stone at summer solstice.

In every culture, celestial bodies were identified with deities. Studying their motions appeared to be only way to learn about divine plans.

Result: astrology – the superstitious belief that events on Earth are governed by events in the sky. Many people still believe this is true.

Eclipses of Sun and Moon, planetary conjunctions, meteor showers, or appearances of comets were usually seen as omens of war or famine.

In every culture, celestial bodies also played role in creation myths. For thousands of years, astronomy was closely related to religion.

First astronomers had one common view: Earth was centre of universe. Go outside, look up and you will understand why.

116. What did ancient civilisations know about the universe?

Not much. Put differently: as much as you would know if you looked around with curiosity but no optical aid or prior knowledge.

Egyptians believed Earth to be flat, with sky goddess Nut bent over Earth god Geb. Every day, Sun god Ra moved over Nut's body in boat.

Bright star Sirius related to goddess Isis. Important for agriculture: first seen in morning sky in June, heralding flooding of Nile.

Pyramids face directions N, E, S & W, but have no observational purpose. Popular idea that ground plan related to Orion is probably false.

Babylonians built temple towers (ziggurats) for religious purposes, but also for observing night sky. They left cuneiform records.

Oldest records: lunar eclipse (before 2000 BC), and 21-year-long observations of Venus (called Ishtar), seen as morning or evening star.

Over centuries, they discovered cyclic nature of sky events, like planetary motions and eclipses. This allowed them to predict them.

Babylonians also left us with division of day into 24 hours, of circle into 360 degrees, and of zodiac into 12 constellations.

In China & Korea, court astrologers kept track of celestial events. Many records remain of comets and 'guest stars' (supernovae).

None of these cultures tried to understand the mechanisms behind heavenly motion. Celestial bodies were seen as ethereal/heavenly.

117. What was the Greek view of the universe?

Greeks knew much more. Thales of Miletus predicted solar eclipse of 28 May 585 BC, which ended war between Medes and Lydians.

~500 BC, Parmenides concluded Earth is sphere. Reason: Earth's shadow during lunar eclipse is always circular. Only sphere can do that.

Math and geometry of Pythagoras and Plato laid basis for Greek world view. Sphere and circle as perfect shapes; important role for numbers.

Plato's student, Aristotle (384–322 BC), came up with idea of Earth surrounded by invisible crystalline spheres carrying celestial bodies.

Aristarchus of Samos (310–230 BC) determined Sun is 19 x further away than Moon. Wrong, but respect for even attempting an answer.

From observing Sun in Alexandria and Syene (Aswan), Eratosthenes of Cyrene (276–194 BC) got pretty close figure for circumference of Earth.

Hipparchus of Nicaea (190–120 BC) discovered slow change in orientation of Earth's axis, and compiled first star catalogue, ~80 stars.

Greek view: Earth surrounded by seven 'planets' (Moon, Mercury, Venus, Sun, Mars, Jupiter, Saturn) and outermost sphere of fixed stars.

Geocentric (Earth-centred) world view was improved/ expanded by Claudius Ptolemaeus (AD Ptolemy, 90–168), who lived/worked in Alexandria.

Ptolemy used epicycles to explain observed complex planetary motions: planet moves on epicycle; empty centre of epicycle orbits Earth.

Also, Earth could be slightly off-centre in circular planetary orbit. Eventually, Ptolemy needed dozens of epicycles and other tricks.

Around AD 150, he published his ideas on planetary motion, eclipses etc. in collection of 13 vols called *Mathematical Treatise*.

Ptolemy's book (aka the *Almagest*) also contained catalogue of 1022 stars and list of 48 constellations, which are still in use today.

118. How did Greek ideas survive the Dark Ages?

Ptolemy's ideas about geocentric universe, with Sun, Moon and planets orbiting central Earth, held sway for 1400 years.

During most of this time, Europe was in intellectual Dark Ages. But Greek heritage was kept alive – and improved – by Arabic astronomers.

Islamic Golden Age (flourishing art and science, mainly at Muslim courts) lasted from 8th century to 1258, when Mongols destroyed Baghdad.

Caliph Harun al-Rashid (763–809) had Greek texts translated into Arabic. Ptolemy's book became known as the *Almagest* ('the Greatest').

Many stars also have Arabic names: Aldebaran (follower), Betelgeuse (armpit of the central one), Deneb (hen's tail), Altair (flying eagle).

Persian astronomer Abd al-Rahman al-Sufi (903–86) published *Book of Fixed Stars*; discovered Andromeda galaxy and Large Magellanic Cloud.

Abu Rayhan al-Biruni (973–1048) was great observer. Invented astronomical instruments, refuted astrology, even suggested Earth orbited Sun.

Abu Ishaq Ibrahim al-Zarqali (1029–87) lived in Moorish Toledo, Spain. Compiled tables to calculate positions of Sun, Moon and planets.

Arabic books from library in Toledo, including Ptolemy's *Almagest*, were translated into Latin around 1175 by Gerard of Cremona and others.

Thus, ancient Arabic, Greek and Jewish texts on astronomy, maths and medicine first became available to European scholars in 12th century.

Meanwhile, Arabic astronomy lived on. Ulugh Beg (1394–1449) built observatory in Samarkand; made very precise measurements with naked eye.

119. Why does the Maya calendar stop in 2012?

First thing to say: Don't panic. World won't end 21 December 2012. Nor will there be colliding planets, giant floods or solar superstorms.

Yes, it's true: Old Maya calendar says grand cycle of 13 baktuns (each 144,000 days) will end on that date. End of the 'fourth world'.

According to old Mayan belief, three earlier worlds have likewise ended before. But present-day Maya (Guatemala) are not concerned at all.

Maya culture peaked ~AD 900; lasted until Spanish Conquest in 16th century. Maya had complex number system and hieroglyph-like writings.

They knew of 'dark constellations' (dust clouds in Milky Way) like Condor, Jaguar and Baby Lama. Temples may have doubled as observatories.

They lived south of Tropic of Cancer, so Sun passed through zenith twice a year. Important calendar dates, like first sighting of Pleiades.

Maya also had great interest in bright planet Venus with its 584-day and 8-year cycles. But they had very little knowledge of astronomy.

Apart from Tzolkin and Haab (religious and civil calendars), they used Long Count calendar for historical purposes. Designed in 5th century.

Long Count: 1 baktun = 20 katun, 1 katun = 20 tun, 1 tun = 18 uinal (360 days), 1 uinal = 20 kin, 1 kin = 1 day. So 1 baktun = 144,000 days.

Current grand cycle of 13 baktuns (~5125 years) started 11 August 3114 BC and will end 21 December 2012. Marks start of new grand cycle.

However, no reason for cosmos to comply with calendar system of one particular culture on Earth. 22 December 2012 is just another Saturday.

120. Who came up with the idea of a Sun-centred universe?

Nicolaus Copernicus (1473–1543) revolutionised astronomy with publication of heliocentric (Sun-centred) world view. Superseded Ptolemy.

But Copernicus wasn't the first. Aristarchus toyed with the idea, as did al-Biruni. Copernicus almost certainly knew of Aristarchus' work.

Niklaus Koppernigk (Polish name) was born in Torun. Father died when he was 10. Raised by his uncle, Bishop Lucas von Watzenrode.

Copernicus studied astronomy, theology, canon law and medicine at the universities of Krakow (Poland), Bologna and Padua (both in Italy).

After returning to Poland, he became canon at Frombork Cathedral in 1497. Enough time to work on his ideas about a Sun-centred universe.

Around 1530, he completed the manuscript of his book, *De Revolutionibus Orbium Coelestium* ('On the revolutions of the celestial spheres').

In 1539, when Copernicus was 66, his 25-year-old pupil Georg Joachim von Lauchen (Rheticus) urged him to publish it, & found a printer.

The book was published in Nuremberg in 1543. According to legend, Copernicus saw the first copy on the day of his death, 24 May.

Copernicus' world view: Sun orbited by Mercury, Venus, Earth, Mars, Jupiter and Saturn. Beyond Saturn is sphere of fixed stars.

Surprisingly, Copernicus still needed dozens of epicycles, just like Ptolemy. Why? He still stuck to Greek idea of perfect, circular orbits.

Grave of Copernicus in Frombork Cathedral was discovered in 2005, announced 2008. On 22 May 2010, he was given a second ceremonial funeral.

121. When did astronomy turn into a true science?

European astronomers embraced Sun-centred world view of Copernicus. But motions of some planets, notably Mars, remained hard to explain.

In 1609, using observations of his tutor Tycho Brahe, Johann Kepler (1571–1630) solved problem: planetary orbits are ellipses, not circles.

Galileo (1564–1642) was first to publish telescopic views of the sky. Phases of Venus and moons of Jupiter supported heliocentric theory.

In 1687, Isaac Newton (1642–1727) published *Philosophiae Naturalis Principia Mathematica*, describing his law of universal gravitation.

Upshot: falling apple is governed by same law as orbiting planets. Newton provided physical basis for Kepler's laws of planetary motion.

18th-C discoveries: periodicity of comets, 'proper' motion of stars on sky, shift in position (aberration) of stars due to Earth's motion.

Bigger telescopes revealed more stars, nebulae, and, on 13 March 1781, a new planet: Uranus, discovered by William Herschel (1738–1822).

First asteroid: 1801. Stellar distances: 1838. Spiral nebulae: 1845. Neptune: 23 September 1846. First solar flare: 1859. Age of discovery.

New tools: photography and spectroscopy (decomposition of starlight). Paved way for 'astrophysics': studying physical properties of stars.

True nature of spiral nebulae (galaxies), expansion of universe, and energy source of Sun and stars were discovered between 1920 and 1940.

Current view: humans part of grand, interconnected universe. We are made of star stuff; without prior cosmic evolution we wouldn't be here.

The Telescope

122. Who invented the telescope?

No one knows for sure. First primitive telescopes may have been around late 16th century, maybe even earlier. Very poor quality, though.

First mention of telescope ('tube to see far') is in patent application dated 25 September 1608, by Dutch spectacle-maker Hans Lipperhey.

Lipperhey born ~1570 in Wesel, Germany. Lived/worked in Middelburg: Dutch harbour town with famous glass industry. Good for his trade.

On 2 October 1608, Lipperhey demonstrated his instrument to Prince Maurits and Dutch States General in The Hague. Prince was excited.

Main reason: war between Dutch Republic and Spanish empire. Telescope on tower might reveal enemy troops from afar. Also useful at sea.

But others also claimed invention: Zacharias Janssen (another spectacle-maker from Middelburg) and scientist Jacob Metius from Alkmaar.

Result: patent was never granted. Thanks to Lipperhey's demonstration, though, word about invention spread rapidly through Europe.

In summer of 1609, English astronomer Thomas Harriot made first telescopic drawings of Moon. Not published; discovered only in 20th century.

Not much later, Italian physicist/astronomer Galileo Galilei heard about Dutch invention. Quickly, he built much better telescopes.

Galileo discovered lunar mountains, sunspots, moons of Jupiter, phases of Venus, 'ears' of Saturn (turned out to be planet's rings), etc.

Galileo's publication of his discoveries in *Sidereus Nuncius* ('Starry Messenger', March 1610) marks birth of modern telescopic astronomy.

Subsequently, telescope was much improved, notably by Johannes Kepler (Germany) and Christiaan Huygens (Netherlands). More discoveries.

123. How does a telescope work?

Telescope literally brings starlight into focus. Lens in eye does same, but telescope collects more light, so image brighter/ more detailed.

First telescopes used concave lens to focus starlight. Light is bent, or 'refracted', by glass, so these telescopes are known as refractors.

Good example: burning-glass. Sunlight is concentrated by lens. When focused, intensity is high enough to ignite paper or shoelace.

In fact, lens creates small image of Sun (or other light source) in its 'focal plane'. Check for yourself with burning-glass and desk lamp.

Telescope lens also creates image of observed object in focal plane. To see image in detail, necessary to use magnifying glass (eyepiece).

So refractor has two main elements: objective lens to focus light, and eyepiece (or ocular) to observe image. Usually at either end of tube.

Disadvantage of refractor: focusing is slightly different for various colours, so stars show coloured fringes (chromatic aberration).

In 1668, Isaac Newton invented reflector. Instead of lens, uses concave mirror as objective to focus starlight, without colour defects.

Telescope mirror is curved like shaving mirror. Also produces image of light source in focal plane. Check for yourself with bathroom lamp.

More advantages of mirror: 1) needs only one perfectly ground surface, 2) can be bigger without sagging because can be supported from back.

Therefore, all large telescopes are reflectors. Largest refractor, with 102-cm lens, was built at Yerkes Observatory near Chicago in 1897.

Small additional flat mirrors may be used for comfortable viewing. But main telescope principle is always: objective + eyepiece (or camera).

Telescope must be mounted for stability and, ideally, be able to track a star as the Earth's rotation causes it to drift across the sky.

Equatorial mount: easy tracking, but bulky. Alt-azimuth mount: compact, but needs computer control for tracking around two axes at once.

124. Why is bigger always better for telescopes?

It's not just masculine mine-is-bigger-than-yours envy. Bigger telescopes (size of lens/mirror) reveal more details and fainter objects.

The pupil, through which light enters your eye, is tiny (5 mm at most). So a star must be bright to provide enough light to trigger retina.

If your pupil was much larger, your eye could collect more starlight and see much fainter stars. Telescope is a bigger pupil.

Think of an empty wine bottle left out in the rain. It takes ages to fill up. Put a funnel in its neck, though, and it fills up quickly.

Big lens or mirror funnels more starlight, so big telescopes can see fainter objects, or, alternatively, same objects much further away.

Bigger scopes can also see finer details (better spatial resolution). Eg. single star, seen with big scope, may turn out to be binary . . .

. . . Or surface details on Moon/Mars. Or substructure (spiral arms, gas clouds, star clusters) in remote galaxy. More detail is always better.

Actually, magnification isn't that important. It tells you how large an object is projected on your retina, not how much detail it reveals.

So, if you want to impress a telescope owner, don't ask 'What's the magnification?' but 'What's the aperture?' (ie: size of lens or mirror).

BTW, turbulence in atmosphere determines the detail a telescope can see. So light-gathering area of mirror is always more important.

10-m Keck Telescope (Hawaii) is ~650 times larger than Galileo's first telescope. Sees 650 x finer detail, and stars over 400,000 x fainter.

125. How do astronomers untwinkle the stars?

To see the stars, you need a cloudless night. But even a crystal-clear sky isn't perfect. Earth's turbulent atmosphere degrades the view.

Starlight passes through moving air bubbles with different temperatures (atmospheric turbulence). Bubbles bend light, like lenses.

Result: stars twinkle, jitter, sparkle, and may even appear to change colour. Nice for romantic lovers; disastrous for astronomers.

No matter how big your telescope, atmospheric turbulence limits resolution to 1 arcsecond at best: equivalent to 5 mm at distance of 1 km.

Surprising fact: decent amateur scope has same resolution as 10-m Keck Telescope. Keck, of course, has much larger light-gathering power.

To untwinkle stars, astronomers use 'adaptive optics'. Idea: measure effects of turbulence and, moment by moment, correct telescope image.

100 times per second, wavefront sensor measures how starlight is affected by turbulence. Fast computer calculates necessary corrections.

Surface of small, 'rubber', mirror, close to focus, can be flexed using piezoelectric crystals (which expand in response to current).

Rubber mirror is made to undulate in precise way needed to cancel out distortions caused by atmosphere. It's like removing the atmosphere!

Using adaptive optics (AO), big telescopes achieve eagle-eyed vision. Nowadays, almost all big telescopes are outfitted with AO.

Sometimes, sodium laser is used to create artificial 'guide star' high up in atmosphere to provide info on atmospheric turbulence.

AO was originally developed by US military: spy satellites also have to look through turbulent atmosphere, albeit down rather than up.

126. Why do astronomers couple together telescopes?

Bigger telescopes provide sharper views of the universe. But you get the same result by coupling two or more smaller telescopes together.

Technique is called interferometry. Trick is to make the detector believe that the 2 telescope mirrors are part of one huge, single mirror.

To understand, imagine mirror with diameter of 100 metres. Would have huge light-gathering power and extremely high resolution.

Painting black spots on mirror reduces light-gathering power. But not resolution, as long as there are still some working parts 100 m apart.

Next, paint all of mirror black, except for 2 circular 10-m patches on opposite ends. Resulting images will be dim, but still very sharp.

Now cut away black parts. That leaves two 10-m scopes, 100 m apart. Coupled together, they have same sharp vision as imaginary giant scope.

Trick works only when detector at focus receives starlight from both scopes 'in phase' – crests/troughs of light waves need to match up.

So with two telescopes on the ground, high-tech 'delay lines' with nanometre precision are needed to let starlight always arrive in phase.

Much less precision is needed for longer wavelengths, eg radio waves. Very Large Array in New Mexico is example of radio interferometer.

Today, interferometry also works for large optical/infrared telescopes. Keck Interferometer couples 2 identical 10-m telescopes, 85 m apart.

Four identical 8.2-m scopes of European Very Large Telescope (Chile) can be coupled to obtain resolving power of a 120-m telescope.

127. What are the largest telescopes on the ground?

As of 2011, there are fourteen groundbased optical telescopes with an aperture of over 8 metres. Six of them are in the southern hemisphere.

The largest is the Gran Telescopio Canarias (GTC) on the Spanish island of La Palma. Its 10.4-m mirror consists of 36 hexagonal segments.

GTC is based on design of twin 10-m Keck telescopes on Mauna Kea, Hawaii, which are operated by two Californian institutions and NASA.

Also on 4200-metre-high Mauna Kea are Japanese 8.3-metre Subaru ('Pleiades') telescope and international 8.1-metre Gemini North telescope.

As name implies, Gemini has twin (Gemini South), at Cerro Pachón in northern Chile. Subaru and Gemini have monolithic (one-piece) mirrors.

Few hundred kilometres north of Pachón is Cerro Paranal (2635 m), home to European Southern Observatory's Very Large Telescope (VLT).

VLT consists of four identical 8.2-m telescopes: Antu, Kueyen, Melipal and Yepun (Sun, Moon, Southern Cross and Sirius in Mapuche language).

On Mt Graham, Arizona, sits the Large Binocular Telescope: two 8.4-m mirrors on the same mount, working together as an interferometer.

Two remaining giant telescopes are Hobby-Eberly Telescope (Mt Fowlkes, Texas) and Southern African Large Telescope (South Africa).

Both have segmented mirrors 11 m across, but, due to their construction, 9–10 m at most is used. They also have limited view of sky.

Remote mountaintop locations of big scopes are important: skies are clear, dry, dark (little light pollution) and quiet (little turbulence).

128. When will the Hubble Space Telescope be replaced?

The Hubble Space Telescope, which is in low-Earth orbit, is named after American cosmologist Edwin Hubble. It was launched in April 1990.

Why space? 1) Sky is black, 24/7. 2) No atmosphere means no turbulence. 3) Above atmosphere, can observe UV and infrared, usually absorbed.

Downside: extremely expensive; hard to service or repair; and, due to launch constraints, quite small – mirror is only 2.4 m across.

Hubble has been serviced five times by shuttle astronauts. They fixed or repaired broken parts and installed new, more sensitive cameras.

As a result, Hubble is now much more powerful than 20 years ago. Fair to say it has revolutionised astronomy. And made terrific pics too.

However, May 2009 servicing mission was the last. Hubble may well live for another 10 years, but if something crucial fails, it will die.

BTW, Hubble won't ever return to Earth intact. After it dies, it will make a controlled descent through atmosphere and plunge into ocean.

Hubble's successor, the James Webb Space Telescope, is being built by NASA. Project is years late and way over budget.

Webb has much larger, 6.5-m, segmented mirror. It will be deployed in space with a big sunshade to protect sensitive mirror/instruments.

Webb won't orbit Earth. Will be placed instead at point in space 1.5 million km away from Earth, in opposite direction from Sun.

Reason: Webb will observe infrared (heat) and so needs to be far away from hot Earth. Launch on European *Ariane* planned for 2018.

129. What will future telescopes look like?

Not very much different from current telescopes. Just bigger. Much, much bigger. At least, that's how they look on the drawing board.

Using spinning ovens, giant telescope mirrors up to 8.4 metres wide can be cast in one piece. For larger apertures, tricks are needed.

One trick – multiple mirrors on one mount – will be used for the Giant Magellan Telescope (GMT), to be built at Cerro Las Campanas, Chile.

The GMT will consist of seven 8.4-m mirrors: six of them surrounding a central, seventh one. Together they have the power of a 24.5-m scope.

Two other planned future telescopes will have segmented mirrors, like Keck. But where Keck has 36 segments, these giants will have hundreds.

The Thirty Meter Telescope (TMT) is an international project led by US and Canada. Planned location is Mauna Kea, Hawaii, close to Keck.

ESO (European Southern Observatory) plans even larger telescope: 39.2 m across. Cerro Armazones in northern Chile, close to Paranal.

ESO already has a Very Large Telescope, so they're calling this one the (European) Extremely Large Telescope. Long live the superlatives.

Being 39.2 m wide, the E-ELT mirror will sport a surface area (hence sensitivity) 70% larger than the Thirty Meter Telescope.

All these giant telescopes of the future are scheduled for completion between 2018 and 2022. If, that is, they get approved and fully funded.

In the distant future, ESO may build a gargantuan 100-m telescope. Yes, they have a name for that, too: the Overwhelmingly Large Telescope.

130. How does a neutrino 'telescope' work?

Neutrinos: subatomic particles from sunlight-generating nuclear reactions. Hold up thumb. 100 million million slicing through every second.

Defining characteristic of neutrinos: antisocial. Not stopped by atoms of normal matter. Nevertheless, they do interact – extremely rarely.

Trick to detect neutrinos: put a large quantity of atoms in their way. This boosts chance that one or two will be stopped.

Neutrino 'telescope' like Super-Kamiokande deep inside mountain in Japanese Alps is 10-storey-high 'baked-bean can' filled with water.

Occasionally, neutrino interacts with proton in water molecule. Subatomic shrapnel in water creates the light equivalent of supersonic bang.

'Cerenkov light' (same as blue light seen in nuclear 'ponds') is picked up by light detectors which line inside of the giant baked-bean can.

Neutrino telescopes have to be deep underground in order to shield them from cosmic ray 'muons', which mimic the signature of neutrinos.

Super-Kamiokande has 'photographed' Sun – at night, not looking up but down through 13,000 km of rock to Sun on other side of the Earth.

Neutrino experiments in Japan and US picked up neutrinos from Supernova 1987A. First neutrinos ever detected from beyond Solar System.

3 types, or 'flavours', of neutrino. Sudbury Neutrino Observatory, Canada, confirmed that, on way from Sun, neutrinos flip between types.

Neutrino 'oscillations' explained puzzling short-fall found by Ray Davis's pioneering 'cleaning fluid' detector. Davis got Nobel Prize.

Newest, most sensitive neutrino telescope: IceCube. Uses 1 km^3 of Antarctic ice as detector. Completed early 2011.

Much excitement about neutrino telescopes. We know what visible universe looks like. But, as yet, no idea what neutrino universe looks like.

Seeing the Universe

131. What is light?

Isaac Newton (1643–1727) thought light consists of tiny particles, travelling in straight lines. Theory described in *Opticks*, 1704.

Christiaan Huygens (1629–95) disagreed. He thought light is a wave, like sound. Theory described in *Treatise on Light*, 1690.

In 1801 in London, Thomas Young showed two light beams can reinforce or cancel each other (interference). Characteristic property of waves.

In 19th C, Michael Faraday and James Clerk Maxwell discovered light is an 'electromagnetic' ripple, crossing space at 300,000 km/s.

Despite apparent wave nature of light, Albert Einstein and Robert Millikan showed light consists of packets, or quanta, of energy (photons).

In quantum physics, light has both particle and wave properties. A photon's energy is related to its wavelength; photons show interference.

Wavelength of visible light ranges from 380 nanometre (violet, high energy) to 780 nm (red, low energy). Sunlight contains all colours.

White sunlight can be dispersed in constituting colours (spectrum) by refraction in droplets of water (rainbow) or by prism.

Rarefied glowing gas only emits at characteristic wavelengths. Sodium lamps: orange glow. Cosmic clouds of hot hydrogen: pinkish glow.

Gases in atmosphere of Sun absorb specific wavelengths. Resulting dark Fraunhofer lines in spectrum carry information on composition.

Polarisation of light gives information about magnetic fields. Red- or blue-shift of spectral lines gives information about motions.

Energy distribution of light (bluer/redder) tells you about temperature of emitting body. All in all, light contains wealth of information.

Visible light is just tiny portion of full electromagnetic spectrum. Astronomers also use instruments to study other types of radiation.

132. What is the speed of light and why is it important?

The speed of light (c) plays role of infinite speed in universe. Just as infinity unattainable, light speed uncatchable by material object.

Why is c unattainable? Energy has mass. If push body faster, boosts energy motion & hence mass. Approaching c, mass skyrockets infinity.

If something is travelling infinitely fast, your speed by comparison negligible. So would appear infinitely fast no matter what your speed.

Similarly, you always measure the same speed of light no matter how fast you are moving relative to the source of light.

Even if someone approaches you at half the speed of light and shines a torch in your eyes, the light strikes you not at 1.5 c but at c.

So that everyone measures same speed (distance/time) for light, everyone's estimates of distance (space) & time must stretch accordingly.

Way in which intervals of space & time appear to change for someone passing depends entirely on how fast they are moving relative to you.

'Moving clocks run slow, moving rulers shrink'. A passing person moves in slow motion & shrinks like pancake in direction of motion.

But 'time dilation' & 'Lorentz contraction' noticeable only if someone moving relative to you at appreciable fraction of speed of light.

Speed of light more than million x faster than passenger jet (300,000 km/s), so effects of 'special relativity' unnoticeable everyday life.

If travelled to stars at close to speed of light, however, time would flow so slowly that, on return, millions of years may have passed.

133. What do radio telescopes listen to?

Radio waves are electromagnetic waves longer than 1 cm wavelength. They constitute the lowest-energy part of the electromagnetic spectrum.

In 1930, Karl Jansky, a radio engineer at Bell Telephone Laboratories, discovered radio waves from the Milky Way. Birth of radio astronomy.

Seven years later, Grote Reber built steerable, parabolic radio dish in his backyard in US. Mapped radio sky; discovered discrete sources.

Advantage of radio astronomy: cosmic radio waves can be studied in bright daylight & even during rain or snow storms – no dark skies needed.

During World War II, Dutch astronomer Henk van de Hulst showed that cold, neutral hydrogen gas should emit faint 21-cm radio waves.

21-cm radiation was first detected in March 1951 by Harold Ewen and Edward Purcell at Harvard, quickly followed by the Dutch in May 1951.

Soon, big radio telescopes were built in Dwingeloo, the Netherlands (1956, 25-m diameter) and at Jodrell Bank, England (1957, 76 m).

Using radio telescopes, astronomers were able to map spiral structure of Milky Way galaxy, and cold, outer regions of other galaxies.

Radio telescopes also pick up 'synchrotron radiation' (at many wavelengths), emitted when fast electrons spiral around magnetic field lines.

Thus, radio astronomy enables study of rapidly spinning pulsars, active galaxies, energetic jets from black holes, and distant quasars.

Largest radio dish: Arecibo, Puerto Rico (305 m), built in bowl-shaped valley. Largest steerable dish: Green Bank, Virginia (100 x 110 m).

Very Large Array (New Mexico) and Westerbork Array (Netherlands) are among largest interferometers: small dishes linked in network.

Future southern-hemisphere Square Kilometre Array (SKA) will be largest radio observatory in history with thousands of small antennae.

134. What does the microwave sky look like?

If you look up at the night sky, you will see scattered stars. But the most striking feature of the night sky is it is mostly black.

Visible light is only tiny portion of 'electromagnetic spectrum'. Other types of (invisible) light include X-rays, infrared & radio waves.

Imagine you have 'magic glasses' and, just by twiddling a knob on the frame, you can change the type of light you see.

If you tune your glasses to X-rays, you will see objects such as black holes. But sky still mostly black. Same for other types of light.

Exception: microwaves, short-wavelength radio waves, the type of 'light' used by mobile phones, TVs and, of course, microwave ovens.

If you tune your glasses to microwaves, the sky will no longer be mostly black. On the contrary, it will be totally, blindingly white.

What you are seeing is the 'afterglow' of the Big Bang fireball. Incredibly, 13.7 bn years after the event, it still pervades all of space.

The 'cosmic background radiation', cooled by expansion of the universe to –270 °C, accounts for 99.9% of all photons in the universe.

Look closely, though. You will see the afterglow is not uniformly white; patches are slightly brighter/slightly less bright than average.

Hot and cold spots in 'afterglow of creation' reveal matter of the Big Bang fireball beginning to curdle into the first-ever galaxies.

The afterglow of the Big Bang shows us universe 380,000 years after its birth. It is the furthest back in time we can see with light.

The fact that the universe – all of space – is still glowing with leftover heat is the most striking evidence that it began in a Big Bang.

135. How do astronomers take the temperature of the universe?

Infrared radiation, with wavelength between 700 nanometre and 1 millimetre, was discovered in 1800 by William Herschel (1738–1822).

Herschel used a prism to create a spectrum of sunlight, from red to blue. He used thermometers to measure energy in each colour.

He noted that a thermometer beyond the red part of the spectrum also received heat, apparently from invisible long-wavelength radiation.

Today, infrared (IR) radiation (heat radiation) is well known, and used in night goggles and consumer camcorders to record night scenes.

In astronomy, cool objects, like dark dust clouds, emit most of their energy at IR wavelengths. IR astronomy reveals the dusty universe.

Dust is also transparent to infrared light. IR telescopes reveal protostars, embedded in dust clouds, even when visible light is absorbed.

Problem: cosmic IR radiation is partly absorbed by water vapour in Earth's atmosphere. Telescope needs to be on high mountain or in space.

Today, most giant ground-based telescopes (like Keck and VLT) are outfitted with both visible-light cameras and near-infrared detectors.

Early IR detectors had no precise directional sensitivity. You couldn't use them to make images of the infrared sky, just blurry pictures.

Now, even your camcorder contains IR-sensitive electronic ccd detectors. Current technology/capabilities comparable to optical detectors.

To be able to 'see' faint IR radiation from space, detectors always need to be cooled close to absolute zero (eg by liquid helium).

First all-sky IR maps produced by IRAS satellite (1983). Found 350,000 sources, including protoplanetary discs and distant dusty galaxies.

Many IR space telescopes followed, like Spitzer Space Telescope (NASA, 2003) and Herschel (ESA, 2009). Hubble also has near-IR camera.

Future 6.5-metre James Webb Space Telescope (NASA/ESA successor to Hubble, launch 2018) will mainly observe at infrared wavelengths.

136. What does the ultraviolet sky look like?

Ultraviolet (UV) light has wavelength between 10 and 400 nanometres. Invisible to human eye, but some animals such as bees can see UV.

UV photons carry much more energy than visible-light photons. That's why UV light from the Sun causes sunburn or even skin cancer.

Luckily, most UV radiation is absorbed in Earth's atmosphere, mainly by ozone. That's why attack of ozone layer by CFC gases was worrisome.

Only very hot objects, like young, massive stars and condensed white dwarfs, emit most of their energy at ultraviolet wavelengths.

Most stars are fainter at UV wavelengths than in visible light. So if we had UV-sensitive eyes, night sky would look quite unimpressive.

Cosmic UV radiation can only be studied from space. Famous UV satellites: International Ultraviolet Explorer (IUE, 1978–96), FUSE (1999).

Hubble Space Telescope also has UV spectrograph/camera, called STIS. Installed in 1997, broke down in 2004, repaired by astronauts in 2009.

Current most active UV space telescope is GALEX (Galaxy Evolution Explorer), launched in 2003. Studies star formation in remote galaxies.

UV telescopes may also reveal presence of warm-hot intergalactic medium (WHIM): very tenuous gas between galaxies and clusters of galaxies.

Oxygen and nitrogen atoms in WHIM, stripped of electrons, reveal presence by absorbing certain UV wavelengths in light from distant quasars.

Meanwhile, UV cameras on board solar space telescopes like SOHO and Solar Dynamics Observatory keep track of explosive flares on Sun.

137. How do astronomers X-ray the universe?

The highest-energy forms of radiation in nature are X-rays (0.01–10 nanometre wavelength) and gamma rays (everything shorter than 0.01 nm).

On Earth, X-rays are used for medical purposes. Energetic enough to pass through human tissue and to cause cancer, if dose is too high.

Gamma rays: even more energetic. Produced in nuclear reactions. Can be lethal. Luckily, Earth's atmosphere blocks cosmic X- and gamma rays.

Rocket experiment in 1949 detected X-rays from Sun. In 1962, another rocket experiment detected first cosmic X-ray source, Scorpius X-1.

Since then, many X-ray satellites have flown, including Chandra (NASA) and XMM-Newton (ESA), both currently operational.

X-rays would go right through telescope mirror, so special optics and/or detectors are needed to take spectra or make images of X-ray sky.

X-rays are produced by extremely hot gas (millions of degrees), eg when drawn into a black hole or shocked in a supernova remnant.

Gamma-ray satellites include the Compton observatory (1991–2000), and Integral (ESA) and Fermi (NASA), both currently operational.

Important field of research: gamma-ray bursts. Most energetic events in universe, caused by exploding giant stars or merging neutron stars.

Mutual annihilation of matter and antimatter, and decay of hypothesised dark matter particles, also produces diffuse gamma rays.

Very high-energy gamma ray photons produce showers of secondary particles in Earth's atmosphere, observable with ground-based instruments.

X-rays and gamma-rays reveal high-energy universe to thrill-seeking astronomers: hottest, most violent and most explosive events in nature.

138. What are cosmic rays?

They're not rays but energetic, charged particles from space, whose origin is poorly understood.

In 1912, flying balloon to 5300 m, Austrian physicist Victor Hess found that atoms in air are stripped of more electrons at higher altitude.

American physicist Robert Millikan wrongly believed such 'ionisation' to be caused by high-energy photons. He coined the term 'cosmic rays'.

About 90% of cosmic ray particles are protons (nuclei of hydrogen atoms); 9% are alpha particles (helium nuclei); 1% are heavier nuclei.

When colliding with air molecules, cosmic rays produce showers of secondary particles and very faint glow known as Cerenkov radiation.

Ground-based particle detectors, spread over large area, register air showers. Ultra-sensitive light detectors register Cerenkov radiation.

Most powerful cosmic ray observatory to date is Pierre Auger Observatory in Argentina: 1600 detectors, distributed over 3000 km^2.

Unfortunately, charged particles are deflected by Milky Way's magnetic field, so arrival direction on Earth is not related to origin.

Ultra-high-energy cosmic rays are protons travelling almost with speed of light, carrying as much energy as well-served tennis ball.

These UHECRs may be 50 million times more energetic than the highest-energy particles produced in any manmade particle accelerators.

UHECRs are very rare. They're not easily deflected. May be produced in relatively nearby active galaxies, harbouring central black holes.

Less energetic cosmic rays are probably accelerated in the blast waves from supernova explosions, but precise mechanism is still unclear.

139. What do cosmic neutrinos reveal about the universe?

Neutrinos are subatomic particles with hardly any mass. They rarely interact with other particles, making them hard to detect.

Neutrinos were postulated in 1930 by Wolfgang Pauli to explain particle experiments. First detection, in a nuclear reactor, came in 1956.

Neutrinos fill the universe. About 400 trillion neutrinos flood through your body each second at light-speed.

Many neutrinos were produced during the Big Bang. Others are produced by nuclear reactions in stellar cores and by supernova explosions.

Neutrinos can be detected by watching large volumes of water: very rarely, they interact with atoms, producing tiny flashes of light.

Detectors are built underground to shield them from cosmic rays. Some large neutrino detectors: Super-Kamiokande (Japan), Sudbury (Canada).

Largest to date is IceCube Neutrino Observatory at South Pole: 1 cubic kilometre of ice, containing thousands of light detectors.

Most neutrinos arriving on Earth come from Sun's core. In 1987, neutrinos from nearby supernova explosion were unexpectedly detected.

While travelling through space, neutrinos change 'flavour' (electron/muon/tau neutrino). Only possible when neutrinos have a small mass.

However, they're so lightweight that relic neutrinos from the Big Bang, despite being very numerous, can't be explanation for dark matter.

Neutrinos are only 'messengers' we receive directly from Sun's core. Study of cosmic neutrinos may also shed light on supernova explosions.

However, main goal of neutrino astronomy is to learn more about fundamental properties of nature, maybe even about mystery of dark matter.

140. What are gravitational waves?

Gravitational waves are hypothetical undulations in the fabric of space-time, travelling with the speed of light, like ripples on a pond.

According to Einstein's general theory of relativity, stiff 4-dimensional fabric of space-time can be deformed/bent by presence of mass.

Likewise, accelerating masses create propagating ripples in space-time that carry away energy. Also called gravitational radiation.

1974: Russell Hulse and Joe Taylor discovered that orbit of binary pulsar B1913+16 is losing energy and shrinking by 3.5 metres/year.

Energy loss is in precise agreement with predictions of general relativity. Binary pulsar apparently emits gravitational waves.

Direct detection is very difficult, however, as wave amplitude is very small. Detectors use laser beams in kilometre-long evacuated pipes.

Even sensitive LIGO detectors in USA have never met with success. 2014 upgrade to higher sensitivity may change that.

Expected sources of gravitational waves: orbiting massive bodies, supernova explosions, gamma-ray bursts, black holes swallowing stars.

Future space-based detectors might also detect high-frequency gravitational waves that are leftovers from the Big Bang.

Gravitational-wave astronomy opens up whole new window on universe. May reveal phenomena never before observed by humans. How cool is that?

Acknowledgements

Govert would like to thank his Twitter followers for their interest in his weekly astronomy 'twourses' on Twitter, which led to this book. And Marcus for his enthusiasm about cooperating on *Tweeting the Universe*.

Marcus would like to thank Neil Belton, Henry Volans, Stephen Page, Felicity Bryan and Karen Chilver for their faith and encouragement. And, of course, Govert.

Also by Marcus Chown

ff

Quantum Theory Cannot Hurt You: A guide to the universe

The entire human race would fit in the volume of a sugar cube. You age faster at the top of a building than at the bottom. Every breath you take contains an atom breathed out by Marilyn Monroe. All of these are true – but why? Two brilliant ideas – quantum theory and Einstein's general theory of relativity – hold the key. Marcus Chown gives us a fascinating, accessible and witty introduction to these two theories that underpin all of modern physics. It's a book you can read in a morning, but which will leave you excited about science for years to come.

'Chown discusses special and general relativity, probability waves, quantum entanglement, gravity and the Big Bang, with humour and beautiful clarity.' *Guardian*

'Readers will experience happy eureka moments.' *The Times*

'A must-read for anyone who wants to better understand this crazy universe we live in. Superb.' *Astronomy Now*

ff

We Need to Talk About Kelvin: What everyday things tell us about the universe

Acclaimed popular science writer Marcus Chown shows how our everyday world reveals profound truths about the ultimate nature of reality. The reflection of your face in a window tells you that the universe at its deepest level is orchestrated by chance. The static on a badly turned TV screen tells you that the universe began in a big bang. And your very existence tells you this may not be the only universe but merely one among an infinity of others, stacked like the pages of a never-ending book . . .

'Perfect for someone who wants a non-intimidating intro to modern physics, or a precocious teenager who won't stop asking why.' *New Scientist*

'Chown writes with ease about some of the most brain-bending of concepts and makes you really think about science.' BBC *Focus Magazine*

'Chown writes very fluently, helping us to visualise things with matchboxes and Lego bricks.' *Guardian*

ff

The Never-Ending Days of Being Dead: Dispatches from the front line of science

Did you ever wonder . . . where did we come from, and what the hell are we doing here? Is Elvis alive and kicking in another space domain? What's beyond the edge of the Universe? Did aliens build the stars? Can we live forever? Acclaimed popular science writer Marcus Chown takes us to the frontier of science, revealing that the questions asked by today's most daring and imaginative scientists are in fact those very ones which keep us up at night. An ambitious yet superbly readable exploration of the mysteries of the Universe.

'We must all be grateful to writers like Chown who are able to make accessible work that in its crude form is not only inaccessible to outsiders, but unknown to them.' *Independent on Sunday*

'I can not find fault with this book, the style is yummy, the mathematics non-existent and the concepts surprising.' *Astronomy Now*

'A limousine among popular-science vehicles.' *Guardian*

ff

Afterglow of Creation: Decoding the message from the beginning of time

It's in the air around you. It carries with it a baby photo of the Universe. Its discoverers mistook it for pigeon droppings yet still won the Nobel Prize. *Afterglow of Creation* tells the story of the biggest cosmological discovery of the last hundred years: the afterglow of the big bang. The result of this find was a 'baby photo' of the universe – sensationally described as 'like seeing the face of God' – which revolutionised our picture of the cosmos. Marcus Chown goes behind the initial hysteria to provide a lively and accessible explanation of this hugely important research – and gives us a compelling and exuberant tale of the human side of science.

'Superbly captures the spirit of scientific discovery.' *Sunday Times*

'A very good piece of storytelling.' *New Scientist*

'A "science for dummies" take on creation, built around an account of the discovery of radiation ripples from the Big Bang.' *The Times*